Cave and Karst Systems of the World

Series editor

James W. LaMoreaux, Tuscaloosa, USA

For further volumes:
http://www.springer.com/series/11987

Martin Knez • Metka Petrič
Tadej Slabe • Stanka Šebela
Editors

The Beka-Ocizla Cave System

Karstological Railway Planning in Slovenia

Co-authors: Franci Gabrovšek, Martin Knez, Janja Kogovšek, Andrej Mihevc, Janez Mulec, Bojan Otoničar, Matija Perne, Metka Petrič, Tanja Pipan, Mitja Prelovšek, Tadej Slabe, Stanka Šebela, Janez Turk, Nadja Zupan Hajna

Springer

Editors
Martin Knez
Metka Petrič
Tadej Slabe
Stanka Šebela
Karst Research Institute
ZRC SAZU
Postojna
Slovenia

Additional material to this book can be downloaded from http://extras.springer.com

ISBN 978-3-319-04455-2 ISBN 978-3-319-04456-9 (eBook)
DOI 10.1007/978-3-319-04456-9
Springer Cham Heidelberg New York Dordrecht London

Library of Congress Control Number: 2014947389

Printed on acid-free paper

Springer is part of Springer Science+Business Media (www.springer.com)

Preface

A good knowledge of the natural and cultural heritage of karst is a precondition for the rational planning of life on it. The karst can be known and understood primarily through the comprehensive study of its surface, caves, waters, and ecological characteristics.

The Karst Research Institute of the Research Centre of the Slovenian Academy of Sciences and Arts has been involved in developing this basic knowledge, establishing interdisciplinary connections among the most important fields of karstology, and in consolidating them into an integral science of karstology for almost seven decades. We try to organize the knowledge to make it as useful as possible for planning life in karst regions and are directly involved in larger major projects. The Institute regularly publishes a selection of new research and knowledge.

Planning without a thorough understanding of the environment and consequently a vision of its development—even though within the boundaries of environmental protection legislation—is certainly not sufficient. We wish to build a foundation for the rational planning of activities on karst based on good karstological research, as much in individual fields of karstology as in interdisciplinary studies. Such planning must take natural and cultural characteristics and vulnerability of karst landscapes into consideration and overcome the inevitable pursuit of profit. Environmental planning must realistically consider the socioeconomic conditions for the benefit of local karst population and the short- and long-term development of karst regions. While the mission of the Karst Research Institute is primarily to expand the basic knowledge of karst, karst phenomena, and karst waters, we are also aware of the need for the continuous and effective communication of karstological knowledge to the wider social community, including through our participation in the more important and directly useful projects.

Ten years ago the Karst Research Institute established a postgraduate karstology program and incorporated karstology courses in the undergraduate geography curriculum. The International Academy of Karst Sciences, an international association of karstologists, was established to link international knowledge and experience more effectively and to find the best foundations for the rational planning of life in various karst regions around the world.

The Karst Research Institute is involved in individual projects, related to the development and protection of the natural and cultural heritage of karst areas, regional planning, water supply systems, the construction of transportation infrastructure, the closure of dump sites in karst areas, the collection of data on karst caves and their protection, karst ecology and determining the extent of human influence on the karst underground, and planning and monitoring the exploitation of karst phenomena for tourism.

We have assembled and published extensive sets of selected directly applicable research studies on karst waters, results of our participation in the planning and construction of expressways on karst, management of caves for tourist purposes, and ecology and protection of the underground. This time we are adding experience acquired in karst planning of the railway route, one of the most demanding projects on Slovenian karst. We recognize that this does not include certain individual topics or the total contribution of karstology to planning life on karst, but we do hope they are a step in the right direction and a challenge for the future.

The area where the new Divača (435 m above sea level)—Port of Koper (0 m above sea level) track will run is divided into two sections. The first leg will run mainly on carbonate rock from Upper Cretaceous, Paleocene, and Eocene. The second leg will run on Eocene flysch. The book presents results of the research on the northern section of the railway running on and through the karst.

The route of the railway near the Divača station on top of the Classical Karst plateau runs about 3 km on the surface and then enters the first, 6,700 m tunnel (T 1), which ends in the upper part of the valley of the Glinščica stream. After a 250 m bridge the railway enters the second, 6,000 m tunnel (T 2). In some places the route runs more than 300 m below the surface.

The last part of the second tunnel passes from the Classical Karst plateau area to the first, 450 m viaduct below the Črni Kal motorway viaduct. The route then runs almost entirely in tunnels T 3, T 4, T 5 and T 6, and in the penultimate T 7 tunnel turns south. With the last, 650 m viaduct the track nears the border with Italy and enters the final, 3,800 m tunnel T 8. The route continues down the Rižana River valley where the otherwise constant 17 ‰ grade eases. The track meets the Port of Koper loading station at sea level.

The total length of the route is just over 27 km, which includes eight tunnels (20.5 km, 75 % of the route), the longest two (near 13,000 m) running through karst. The route also includes three bridges and two viaducts with a combined length of 1,100 m (5 % of the length of the route).

During the initial planning stages, the northern section of the route of the future railway across the karst was considered part of the fifth Pan-European Transport Corridor Barcelona–Kiev.

Martin Knez
Metka Petrič
Tadej Slabe
Stanka Šebela

Acknowledgments

We would like to thank reviewers academician Dr. Andrej Kranjc and Dr. Robert Šajn, translator Lena Djokić, technical and language reviewer Alenka Možina and artist Tamara Korošec.

Contents

Geological Setting

The geomorphological and speleogenetic history of the investigated area depends to a great extent on the geodynamics of the marginal parts of the fold and thrust belt of the Dinaric orogene (i.e. External Dinarides) or the underthrusting of the Istrian Peninsula below the Dinaric range, especially from the Middle Miocene on (Placer et al. 2010). In the process, mainly sedimentary rocks of the Mesozoic passive margin of the Adria-Apulian microplate (Stampfli and Mosar 1999) and Paleogene synorogenic depositional areas have been included. The underthrusting of Istria is a result of the anticlockwise rotation of the Adria microplate (Stampfli et al. 1998) and the related, generally N–S-oriented stresses (Marton et al. 1995; Bressan et al. 1998; Placer et al. 2010), which also caused reactivation and horizontal movements of the older NE–SW striking faults (Jurkovšek et al. 1996; Bressan et al. 1998).

Tectonically, the surroundings of the Glinščica valley with the discussed cave system lie at the border between the Istria–Friuli underthrust zone and the Čičarija anticlinorium (Placer 2005; Placer et al. 2010) (Fig. 1). The latter is a part of the Komen thrust sheet (Placer 1981, 1998) of the Kras–Notranjsko folded structure (i.e. a part of the External Dinaric imbricated belt of the northwestern External Dinarides). These tectonic units correspond to the northwestern part of the Cretaceous Adriatic carbonate platform and the Upper Cretaceous to Eocene synorogenic carbonate platform, which occupied the northeastern part of the Adriatic microplate. Shallow-marine carbonate successions of different Cretaceous formations are separated from the Upper Cretaceous/Lower Palaeogene to Eocene palustrine and shallow-marine limestone of the Kras Group (Košir 2003) by a regional unconformity (Fig. 2). The synorogenic carbonate platform was buried by prograding hemipelagic marl and deep-water clastics (flysch).

The oldest carbonate rocks of the area that crop out in the nearby Matarsko Podolje lowland (Čičarija anticlinorium) comprise Lower Cretaceous limestone and dolostone (Šikić et al. 1972). They are overlain by an Albian to Upper Santonian shallow-marine carbonate succession composed predominantly of limestone with local intercalations of dolostone and coarse-grained non-depositional calcareous-dolomitic breccia. The sequence ends with a second-order unconformity marked by a prominent paleokarstic surface (Otoničar 2007), which is overlain by up to a few tens of metres thick sequence of Upper Cretaceous (Maastrichtian) dark grey palustrine to Lower Eocene shallow-marine foraminiferal limestone that gradually passes over to hemipelagic marl and deep-marine siliciclastic flysch. The unconformity typically separates the underlying passive margin carbonate succession from the overlying deposits of the synorogenic carbonate platform at the periphery of the foreland basin (Košir and Otoničar 2001; Otoničar 2007).

The Cretaceous and Palaeogene sequences of the wider vicinity of the investigated area are parts of the southwestern flank of the aforementioned Čičarija anticlinorium (Placer 1981) and the imbricated structure of the Istria–Friuli underthrust zone with an alternate dip of the strata. The area is dissected by numerous faults with two main directions, NW–SE and NE–SW (Placer 1981, 2005; Placer et al. 2010; Jurkovšek et al. 1996). The most important NW–SE striking faults have a mainly thrust fault character, while transverse to this direction normal and strike slip faults are found. The folds and faults display the multi-phase kinematic evolution of the

M. Knez et al. (eds.), *The Beka-Ocizla Cave System*, Cave and Karst Systems of the World,
DOI: 10.1007/978-3-319-04456-9_1, © Springer International Publishing Switzerland 2015

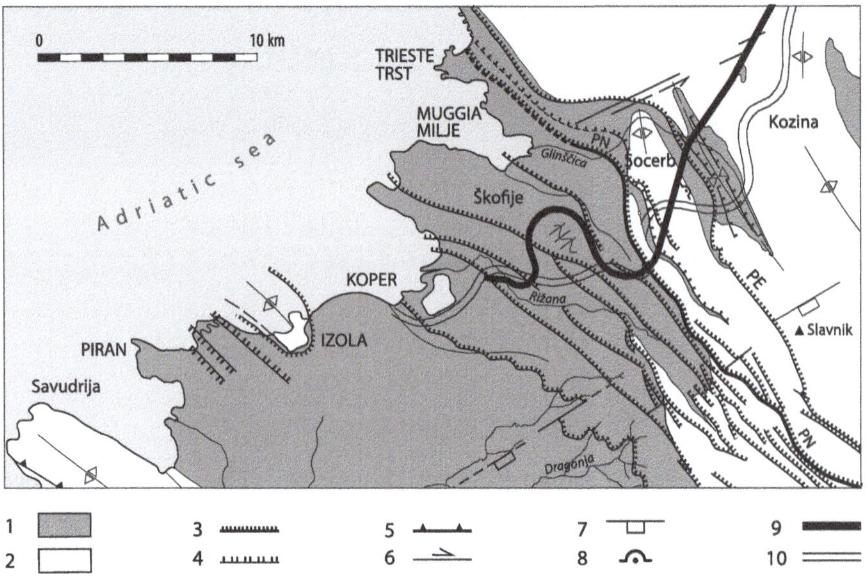

Fig. 1 Istria-Friuli underthrust zone. *1* upper ductile horizon: flysch, *2* platform carbonates, *3* thrust faults (*PE* Petrinje thrust fault, *PN* Palmanova thrust fault), *4* secondary thrust faults of the Strunjan structure, *5* thrust front of External Dinarides, *6* strike-slipe fault, *7* normal fault, *8* Cave, *9* railway, *10* motorway (from Placer et al. 2010)

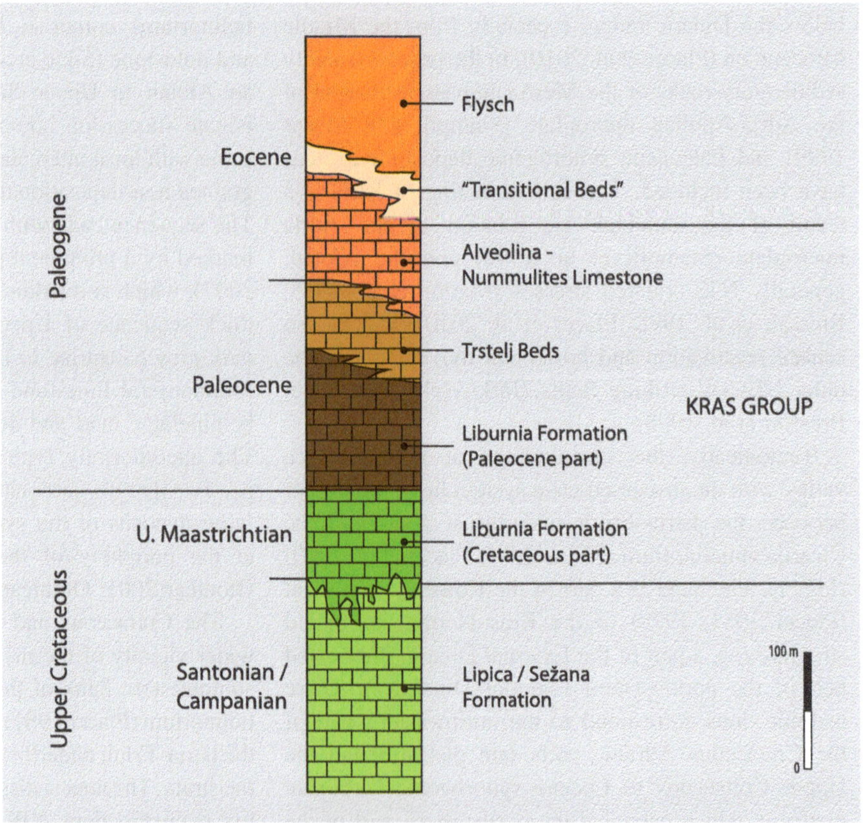

Fig. 2 Generalized Upper Cretaceous to Eocene lithostratigraphic column of the Kras and Matarsko Podolje, SW Slovenia (modified from Košir 2003)

area, which is related to the changing stress fields from Cretaceous to the recent (Jurkovšek et al. 1996). The Beka-Ocizla cave system is located in the highest major slice of the Istria–Friuli underthrust zone, which is determined by two major thrust faults of the zone, i.e. the Palmanova fault and the Petrinje fault (Fig. 1).

In the area of the Glinščica valley, the position of the caves depends on the alternation of carbonate and siliciclastic rocks (Fig. 1), while the internal morphology of the cave passages is also controlled by numerous structural unconformities, such as faults, joints and bedding plane partings. The entrances to the discussed cave system and the morphology of its inner parts are also strongly controlled by the contact between carbonate and non-carbonate lithologies and the numerous joints related to the aforementioned nearby faults (see Chaps. 3 and 5). The entrances to the cave system are related to streams that flow from the northeasterly located slice of siliciclastic flysch, then pass over the so-called transitional beds, i.e. hemipelagic marl and marly limestone (Jurkovšek et al. 1996), and sink at the contact with Eocene limestone. Collapses commonly occur in the transitional beds.

Karst Surface

The formation of the karst surface is a result of surface dissolution of rocks, removal of the rock in form of solution and of the underground flow of water through the voids caused by corrosion, i.e., caves. Thus, the relief shows the organisation and form of the underground drainage system or the type, size and distribution of cavities, which are otherwise not accessible or known to men.

Due to tectonic raising of the entire Classical Karst and geomorphological development of the surface and underground, the underground drainage paths changed in the past; new paths were formed, while the old, now inactive caves were preserved. By studying the current hydrological conditions, old cavities cannot be detected because they were formed under different hydrological conditions. The existence of such caves can be deduced from the relief forms of the surface and from the geomorphological development of the relief.

The area of the railway route contains fluvial relief on flysch rocks and karst relief on limestone. From the relief forms and from the impact of the fluvial relief on the karst, we can deduce the geomorphological development of the entire territory and indirectly also the development of the karst underground or distribution of karst porosity.

We have mapped and analysed the characteristic or diagnostic karst forms on the surface above the first two tunnels, and studied the broader area where development of the relief is connected with development of the karst on the tunnel route. We focused our attention primarily on the location of larger, flat segments of the surface, and on the size and distribution of dolines and collapse dolines.

Special attention was devoted to mapping the caves, which the karst denudation had opened up to the surface. Today, these are hydrologically inactive caves that bear witness to the oldest developmental stages of the karst. Observations and measurements have shown that the greater part of dissolution of karst rocks, around 90 %, takes places in the upper metres of the karst or on the very surface of the karst. This results in the surface of the Classical Karst being lowered by some 20–60 m per million years. As the surface is being lowered, it cuts through numerous old cavities that thus become a part of the present karst relief. Such caves were compared with the current, known caves and in Podgorski kras also with the density of the caves discovered at the Črnotiče quarry.

In addition to the individual relief forms, larger groups of forms that are connected to a special geological structure and share common morphological development features are also important. Such morphostructural or morphogenetic karst units on the route of the tunnels are Divaški kras, Gradišče, the edge of Matarsko podolje lowland, and Podgorski kras or Petrinjski kras (Fig. 1).

1 Geomorphological Analysis of the Territory

1.1 Relief Forms and Groups of Relief Forms

1.1.1 Small Corrosive Forms

Corrosion features of various sizes are formed on the surface of limestone and create an uneven and coarse surface of the rock. These forms are mostly rain flutes and karren. With their help we can deduce the solubility of the rock and the speed of the corrosion of the surface.

M. Knez et al. (eds.), *The Beka-Ocizla Cave System*, Cave and Karst Systems of the World,
DOI: 10.1007/978-3-319-04456-9_2, © Springer International Publishing Switzerland 2015

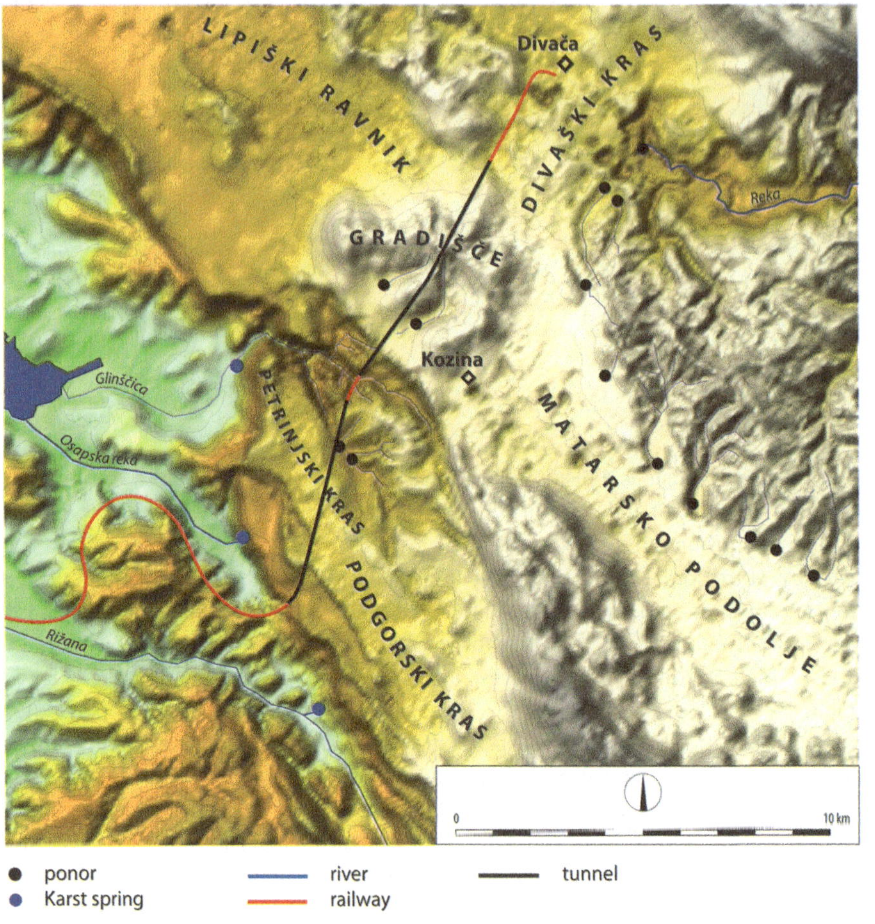

Fig. 1 Digital elevation model (DEM). Areas of the planned railway line and tunnels *T1* and *T2*. Larger relief units have been marked

1.1.2 Karst Hollows

Dolines are the most common karst depressions. They are funnel-shaped or kettle-shaped, mostly up to 10 m deep and with a diameter of up to 50 m. Based on their formation, dolines are divided into corrosion dolines, collapse dolines and dolines that were formed from cavities by denudation.

Solution dolines are formed where dissolution of rock is the strongest and where percolation into the karstified rock is possible. Solution dolines are formed by the locally concentrated dissolution of limestone, due to thicker soil. A larger quantity of CO_2 dissolved in water dissolves more limestone beneath the soil, is released in it, and thus a hollow is formed. These dolines have solid, compact floors beneath them. The majority of the dolines on the route are of this type.

Dolines are most commonly found on karst flatlands and very rarely on slopes or not at all on steeper slopes. These forms are a surface phenomenon; the only condition for their formation is draining of torrential water through the karst (Fig. 2).

When studying dolines in a specific territory, we first have to genetically separate different types of dolines and then, based on them, see the type of karstification.

Collapse dolines are large depressions that were formed with slow collapsing of roofs above larger underground cavities and afterwards with

Fig. 2 A typical doline in the karst of Divaški kras near the location where the tunnel *T1* will commence

transformation of slopes. They are from 50 to 200 m deep and up to 500 m wide, while their volume amounts to several million m^3. Seeing that their dimensions are much greater than dimensions of the largest known halls, and since the majority of the collapse dolines are located near larger underground flows, it seems that the formation of collapse dolines is not only the result of the collapse of the karst cavity, but also above all the result of dissolution and carrying away of the collapsed rock in the depths. The study of collapse dolines and large collapse halls near the Škocjanske jame caves and in the Kačna jama cave near Divača has shown the dual role of karst water in the formation of the collapse dolines (Fig. 3).

Fluctuation of water in the underground, for example over 100 m in the karst area of Divaški kras, causes the water to flow from the main channels into the side channels and into the tectonically fractured zones as the flood rises. When the water level drops, the water withdraws from the fissures. This constant alternation of water causes substantial corrosion and enables formation of rock instability in the depths. Another important role of the proximity of the main waterways is the carrying away of collapsed rock in the form of a solution or partially in the form of particles. The collapse dolines, therefore, point to the proximity of larger water flows.

The surface area of collapse dolines in the karst area of Divaški kras comprises around 6 % of the entire surface area. Their volume is substantial. Near Divača, the large valleys of Risnik, Radvanj and Bukovnik are located. Their bottoms are situated between 360 and 380 m above sea level. Beneath the Bukovnik valley, there are the lower passages of Kačna jama; directly beneath it, one of the passages ends in a large collapse. Here, the river is located 154 m above sea level or 230 m beneath the bottom of the collapse doline.

Favourable conditions for the formation of collapse dolines in the vicinity of the planned route were found only in the karst of Divaški kras.

Collapses that would lead to collapse dolines were not detected from the morphology anywhere along the route. However, collapses may appear where the tunnels will run close to an underground cavity.

1.2 Karst Levelled Surfaces

Large, flat surfaces can be formed only where a greater part of the corrosion ability of the water, precipitation

Fig. 3 Risnik collapse doline (Divača is in the background). The surface is levelled at around 420 m and the *bottom* of the collapse doline at 365 m *above* sea level

or sinking streams is used upon the surface. This can occur when the karst is situated at a low altitude or dammed up with impermeable or poorly permeable rock. Such a position causes the level of the karst water to be located shallowly beneath the surface; the low gradient creates branched out cave systems.

Numerous flat surfaces have been formed on the study area. The main are Divaški kras, Matarsko podolje and Podgorski kras. When afterwards the level of the karst water dropped by several hundred metres, the surface became dissected into numerous dolines, yet preserved the basic flatland shape.

1.3 Contact Karst

Contact karst is karst formed at the contact between surface and underground drainage. The quantity, regime and sediments of the allogenic rivers that flow onto the karst have modified the karst processes. The result of this are special relief forms, e.g. blind valleys and border flatlands, and phenomena such as sinking of water, floods and depositing of sediments onto the karst.

The development of cavities in the karst, which is passed by the tunnel route was influenced by the contact karst along the edge of the Brkini hills, the contact karst of Ocizla and the contact karst near Gročana and Vrhpolje.

Characteristic for the contact karst beneath the Brkini hills are the large levelled surface Divaški kras next to the ponors of the Reka River and the large blind valleys along the lowland Matarsko podolje, which have similarly shaped bottoms and slopes.

The surface forms of the contact karst of the Reka River are limited mainly to the Vremska dolina valley where a large blind valley was formed at around 370 m above sea level.

The bottoms of the blind valleys along the edge of Matarsko podolje are situated at 490–520 m above sea level. Across a span of some 20 km, the bottoms of the blind valleys vary by several tens of metres, while the system of valleys is raised by 200 m. These heights undoubtedly point to a common level, to which deepening and development of blind valleys along the edge of Matarsko podolje were carried out.

The contact karst near Vrhpolje was formed along the edge of a small patch of Eocene flysch, which enables draining away on the surface. Along the edge of the flysch, beneath Vrhpolje, at the altitude of some 450 m, and near Gročana, two streams sink into small indistinct sinkholes.

The contact karst along the Ocizla sinking streams is a large, elongated relief depression with the characteristic name Loke, which continues above a pass, only 30 m higher, into a deeply incised fluvial valley of Griža, a tributary of the Glinščica River. The shape of the depression shows that the stream initially flowed on top of the contact and did not sink. When the stream began to sink, its erosion ability decreased, and the deepening of the valley stopped at an altitude of some 380 m. Afterwards, the top part of the valley was drained away only karstically and became lowered to the level of the present ponors, whereas the non-karstic part of the Botač valley continues to be incised intensively. The developmental stages are also expressed in the morphology of the cave passages of the Ocizla sinking streams.

From the flysch ridge between Beka and Ocizla villages three larger streams and one smaller stream flow into Loke, which was formed along the contact between limestone and flysch. The Ocizla sinking streams drain water from a 3.5 km^2 catchment area on the flysch. The streams flow in from the watershed at the altitude of up to 500 m, while the ponors are situated at the altitude of around 350 m.

Due to the high gradient in the karst, the streams do not flood in front of ponors and do not deposit sediments.

1.4 Unroofed Caves

An unroofed cave is a cavity that has been opened up to the surface by dissolution and lowering of the surface (Fig. 4). Karst caves that were formed deep beneath the surface, usually no longer change after the water course abandons them. However, the surface does change; due to karst denudation, it constantly and evenly continues to lower. The surface lowering rate in the area of the karst is said to range from 20 to 60 m per million years. If presupposing that there were at least 100 m of the roof above the currently denuded caves, then these caves are from 2 to 10 million years old. These ages are confirmed by palaeomagnetic dating in the caves in the Črnotiče quarry and near Kozina.

If a cave had been filled with sediments, it is often not detected in the relief morphology; it is identified only by the typical sediments, mainly by flowstone (Fig. 5), lesser rockiness of the surface where the cave passages have been unroofed and by different use of land in the past, as they were often connected with smaller, elongated patches of arable land. The shape of unroofed caves depends on the size and shape of the cavity (passage, oblique passage) and on the incline of the surface (flat surface, slope), which had cut through the cave.

The findings of denuded remains of the caves on the surface testify to the perforation of the karst segment at this altitude. The dimensions and sediments found in the unroofed caves can lead to conclusions regarding the palaeographic conditions under which they had been formed. We can determine the flow rate through the cave and the sediments can tell us whether a sinking stream ran through it and where its surface catchment area was. Subsequent sediments, flowstone, red loam, rubble and remains of collapses bear witness to the subsequent dry stage of the cave's development and its disintegration, as it was reached by the surface's denudation front.

There have been 17 larger unroofed caves identified in the karst area of Divaški kras. The biggest one, the unroofed cave in Lipove doline, is situated between 420 and 440 m above sea level.

It is filled up with flowstone, gravel and sand, which had been deposited in the cave by a sinking stream, a predecessor of today's Reka River. The cave had passages with sections up to 20 × 20 m.

Near the route lies an unroofed cave, a continuation of the Divaška jama cave. Several unroofed caves were also found on the route between Divača and Lokev.

Less explored are the unroofed caves in Matarsko podolje lowland where several of them were explored during the construction works for the motorway north of Divača.

The highest density of unroofed caves is on the surface above the tunnel on the karst of Petrinjski kras between the Beka-Ocizla cave system sinking streams and the Črnotiče quarry (Fig. 6). Most of the caves have been filled up with loam, sand and gravel; quite a lot of flowstone has also been deposited there. In relief they are shown as shallow, yet long, elongated, 4–8 m wide depressions. They connect dolines, which means that the dolines were likewise formed by transformation of caves in the area of the surface denudation front.

A similar density of unroofed caves and similar surface forms continue in the Črnotiče quarry where we were also able to observe the cross-sections of 16 such caves and study the sediments in them (Fig. 7).

In the quarry we monitored the excavation of a 180 m long cave with a profile of 15 × 15 m. The cave was entirely filled up with massive flowstone, loam and quartz sand. These sediments have not been cemented. The cave and its sediments reached the surface where it was detected as an unroofed cave prior to its destruction.

In addition to horizontal caves, five vertical shafts were also opened up during the work in the quarry. Shafts are more difficult to detect because they were immediately filled up during the work in the quarry. The biggest shaft had a profile of 10 × 8 m and was around 60 m deep. The shaft could not be explored because the area around the entrance to the shaft was

Fig. 4 During the construction of the motorway numerous unroofed caves were discovered near Divača, i.e. caves above which limestone has already dissolved. The proof that they are caves, are the remains of flowstone and stalagmites, and stalactites found inside them, in addition to the other sediments deposited underground by the former sinking stream

heavily blasted. The bottom of this shaft reached the altitude of around 340 m above sea level (Fig. 8).

2 Morphostructural Units

In geomorphology this term denotes larger relief units, the scope and size of which are undoubtedly connected with geological structures. In the terrain in question, they are mostly contacts of various rocks, while the units are often limited with distinct neotectonic lines. The relief forms of one morphostructural unit usually underwent a common geomorphological development, which is why the term morphogenetic units can also be used to describe them.

The morphostructural or morphogenetic units usually exhibit similar relief forms, which is why they often have their own names. The tunnels' route passes through the karst area of Divaški kras, Gradišče and the edge of Matarsko podolje, it crosses the Glinščica valley and runs across the karst of Podgorski kras. For deeper understanding, the morphostructural units of the Brkini hills, the Slavnik mountain and the Taborski griči hills are also important (Fig. 9).

2.1 Divaški Kras

Divaški kras is a lower and rather flat karst plain, situated between the Vremska dolina valley, where the Reka River sinks, and the north edge of the Brkini, Taborski hribi and Gradišče hills. The surface of Divaški kras is located at the altitude between 450 and 400 m. It is more or less flat, with numerous dolines cut into it. Despite the relatively high number and density of dolines, the flat surface cover over 80 % of the total surface area, whereas the dolines cover a small portion, around 6 % (Fig. 10).

The tunnel route will contain a greater number of dolines, i.e. 15 dolines with non-load-bearing sediments at the bottom. Some of these dolines might have been formed from shafts filled with sediments. The problem of dolines is the low load-bearing capacity of the floor, which is why all the sediments must be removed, so that the base of the route lies on rock.

Collapse dolines were formed near the planned route only in the karst area of Divaški kras. Near Divača the large valleys of Risnik, Radvanj and Bukovnik are located. The volume of the collapse dolines is large. Risnik has around 1,700,000 m^3 and Radvanj

Fig. 5 A stalagmite in an unroofed cave in the Lipove doline near Divača. The stalagmite is standing in the open because the rock formation above the cave and the cave roof were dissolved by precipitation water

around 9,000,000 m^3 of volume. The connection between the large collapse dolines and the underground flow of the Reka River is evident. Their bottoms are situated between 360 and 380 m above sea level. Beneath the Bukovnik valley, there are the lower passages of the cave Kačna jama; directly beneath it, one of the passages ends in a large collapse.

The large collapse dolines were formed when roofs caved in over larger underground cavities or underground channels of larger rivers, which carried away the crumbled fragments of rocks. Such is the Radvanj collapse doline, which is located to the east, beyond the route (Fig. 11); however, it indicates that we can expect large passages, perhaps even large halls, at around 350 m above sea level. This should not affect the route as it is located deep beneath the surface across which the railway line will run.

Several larger unroofed caves have been identified in the karst area of Divaški kras. Most of the unroofed caves were discovered during the works on the route of the road near Divača, Dolnje Ležeče and near Povirje. Near the route lies an unroofed cave, a continuation of the cave Divaška jama.

The route of the railway line crosses only one large unroofed cave. The latter is located in the northeastern continuation of Divaška jama. This cave will probably be manifested in the transverse profile as a cavity up to 20 × 20 m large and filled with sediments with poor geomechanical properties (Fig. 12).

2.2 Lipiški Ravnik

Lipiški ravnik is an expansive, flat surface between Lokev, Sežana and Orlek. The surface of the plain is situated at the altitudes between 450 m near Lokev and 400 m near Sežana. The surface is finely dissected with numerous dolines.

The route of the railway line runs across only a section of this plain. Collapse dolines are no longer present here; there is only one unroofed cave on the surface, which partially reaches into the route area at 3.5 km. It is a cave passage filled with sediments of poorer mechanical properties (Fig. 13).

2.3 Gradišče

It is a vast elevation between Matarsko podolje, the Lipiški ravnik plain and the karst of Divaški kras (Fig. 14). It consists of a series of peaks, the highest one being Veliko gradišče (741 m) (Fig. 15). It is made up of Cretaceous and Palaeogene limestone; there is also a patch of Eocene flysch on top, which enables draining away of water on the surface. That is why, in addition to the karst relief, a fluvial relief was also formed on top of it; small brooks flow towards Gročana and Vrhpolje where they sink.

Beneath Vrhpolje, a valley with a flat bottom was formed, which resembles a blind valley. There a brook sinks from the flysch into indistinct holes; during

Fig. 6 Workers in the Črnotiče quarry often come across large caves; this can be seen from the pieces of stalagmites and stalactites which they treat as tailings

higher water levels it flows across the surface to the Glinščica valley.

In the area of the village Krvavi potok or the valley Vrhpoljska dolina (from 7.5 km to the contact with the flysch at 8.8 km), the tunnel T 1 runs around 150 m beneath the surface. The surface, the edge of Matarsko podolje, is levelled at around 450 m above sea level and dissected with only a few dolines. They cannot be used for deducing the underground conditions in this part of the karst (Fig. 16). Two unroofed caves on the surface nearby point to the existence of cave passages, up to several metres wide and high, on the current surface; therefore, one can expect such passages even deeper, at the level of the tunnel.

Fig. 7 Remains of a cave in the wall of the Črnotiče quarry. The passage had a 15 × 15 m profile and was partially filled with loam, sand and flowstone. In the photograph these sediments can be easily distinguished from the surrounding limestone

2.4 The Glinščica Valley

The Glinščica valley was formed in flysch rock between the foothills of the Slavnik mountain and the edge of Matarsko podolje and the karst of Petrinjski kras or Podgorski kras. The Glinščica River gathers the surface inflows on flysch of the karst of Podgorski kras near Podklanec and then deepens into two narrow valleys into flysch and limestone. This valley was formed after the neotectonic sink of the Gulf of Trieste had been formed, into which the valley opens up near Boljunec.

2.5 Podgorski Kras or Petrinjski Kras

It is a distinct karst plateau at the foothills of the Slavnik mountain. The western edge of the plateau drops down into the valley of the Rižana River and Osapska reka River in two sharp folds formed at the overthrust of limestone on flysch. Towards the north it is sharply limited by steep slopes above the Glinščica valley or the neotectonic sink of the Gulf of Trieste above the town of Dolina.

The karst area of Podgorski kras is made up of several larger and smaller overthrusting slices of flysch and limestone. The biggest slice of limestone is the 2 km wide karst area of Petrinjski kras between the flysch of Ocizla and the flysch slice of the Karst Edge between Kastelec and Črnotiče (Fig. 17). The planned route runs across it from the contact with the flysch beneath Ocizla to the exit of the tunnel at Črni Kal. The flysch slices were of great importance in the past development of the karst area of Podgorski kras. They dammed the water in the limestone and enabled the levelling of the surface. Today, the flysch lenses contribute to the disintegration of the plateau; the Glinščica River, which had cut shallow valleys into it, collects water on top of them and at the edge deepens into a deep, eroded valley. The sinking streams of Ocizla also collect water on top of the flysch belts.

Doline-covered flatlands were formed on limestone, while a fluvial relief was formed on flysch. A segment of flysch flows off on the surface into the Glinščica valley; due to the high gradient, it formed deep valleys.

The karst area of Podgorski kras or Petrinjski kras is a large flatland at the altitudes between 400 and 450 m. The tunnel pipes run across a section of it, between 11.5 and 16.0 km. Petrinjski kras is a 2 km wide section of Podgorski kras (Fig. 18). It is situated between the flysch of the Beka-Ocizla streams and the flysch slice of the Karst Edge near Črni Kal. The surface of the karst of Petrinjski kras is flat; small, rare

Fig. 8 A cut cave passage, an unroofed cave in the wall of the Črnotiče quarry. The passage was filled up with loam and sand by the sinking stream, while a 7 m *thick* bed of flowstone and stalactites was deposited on *top* of it. In the section, the passage is 17 m high and up to 5 m wide. The *light green dot* and the letter *M* mark the spot where 2-million-year-old mammal fossils were found in the sediment. Such passages, the unroofed caves, are common in the karst of Petrinjski kras

dolines exceed 50 m in width and 10 m in depth. Also common are very shallow dolines, up to 5 m deep.

The density of the dolines is relatively low, under 10 per km^2. The dolines differ with regard to their genesis. In the Črnotiče quarry, where we have been monitoring the progress of the quarry for several years, we have observed the excavation of several dolines. It has been shown that the majority of them were formed from the former caves (Fig. 19).

The tunnel runs beneath the karst of Petrinjski kras between 12.0 and 15.0 km. Unroofed caves are typical of this entire surface. In this section, the tunnel is located around 200 m below the surface. Although it cannot be precisely determined whether there is a

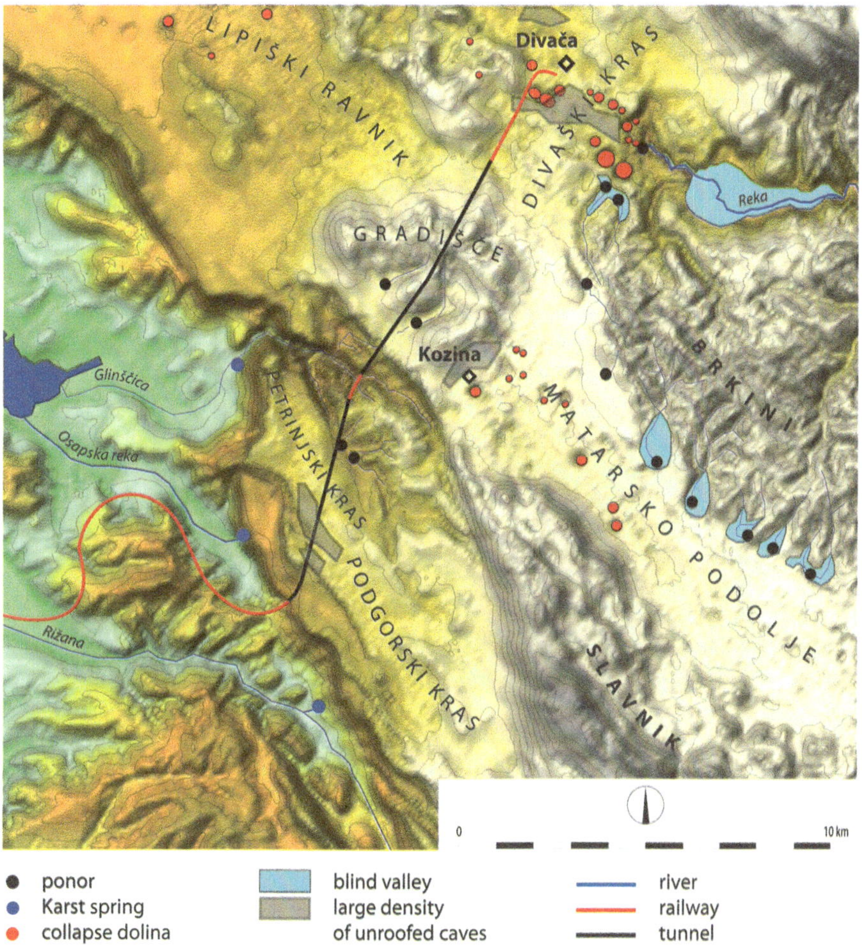

Fig. 9 Morphostructural units in the tunnels' *T1–T2* impact area. Divaški kras, 0.0–3.5 km

similar density of caves in the depths, it is very likely that such a density of cavities accompanies the contact of flysch and limestone also at the height of the tunnel.

2.6 Matarsko Podolje

The Matarsko podolje lowland is a 20 km long and 2–5 km wide flatland between the flysch Brkini hills and the Slavnik mountain chain. Its doline-covered and curved bottom rises steadily from around 490 m at Kozina to 690 m at Starod. At its edge, 17 separate streams flow in from the Brkini and sink into the edge of the lowland. The bottom of the lowland is more or less flat and finely dissected with dolines, collapse dolines and blind valleys.

The lowland and the bottoms of the blind valleys were formed near the level of groundwater; the depth of the accessible caves shows that today the water table of karst water is at least 50–100 m beneath the bottom of the lowland. The bottom of the Jazbina pri Materiji cave is dry and situated 350 m above sea level. The water table of karst water is probably much lower, at around 300 m. The tunnel route runs along the outer northwest edge of Matarsko podolje by the foothills of the Gradišče mountain.

3 Development of the Karst Relief

From individual relief forms, and primarily from the relief groups, we can reconstruct the development of the relief of the entire area of the south edge of the

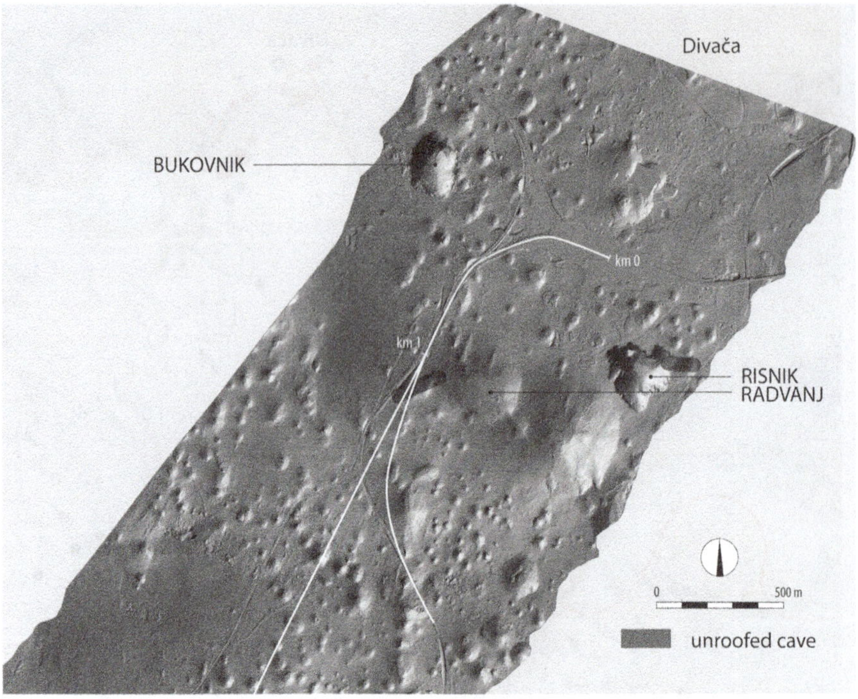

Fig. 10 DEM of the surface of Divaški kras. The large collapse dolines of Bukovnik, Risnik and Radvanj stand out. Dolines are small and evenly distributed across the flat relief. Larger unroofed caves can be found only to the west of Radvanj. A large series of dolines in the N–S direction was formed along the important fissured zone

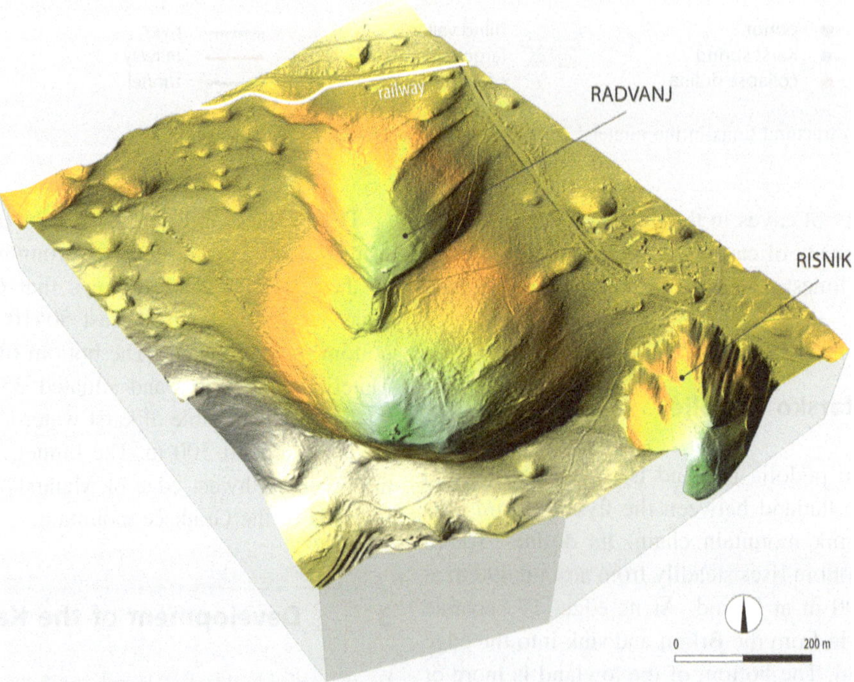

Fig. 11 The double collapse doline of Radvanj and Risnik near Divača. The railway route runs beyond the area of the collapse doline, merely crossing a larger unroofed cave

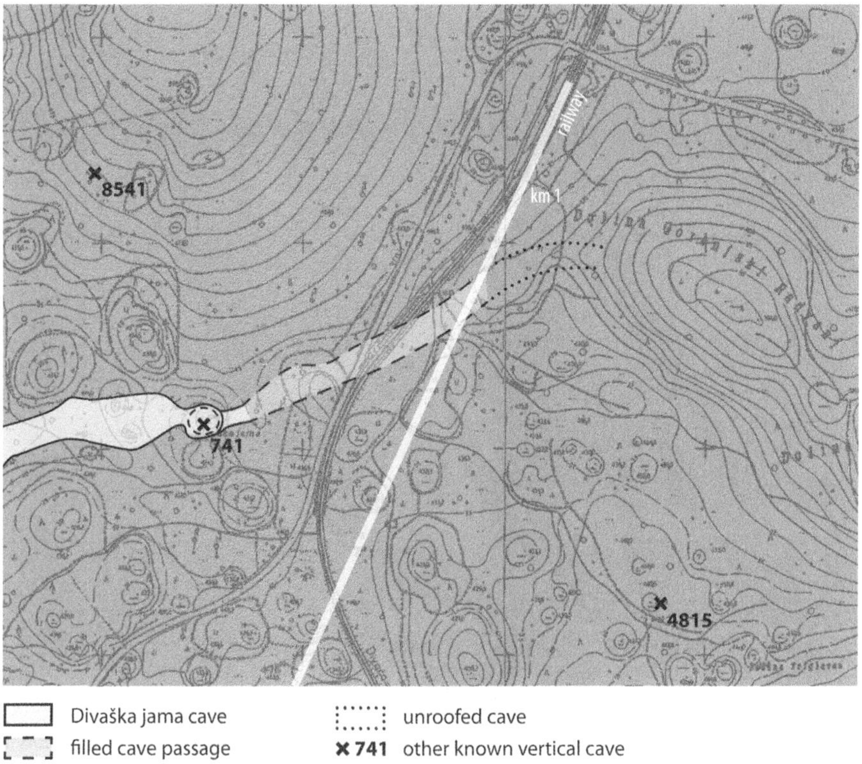

Fig. 12 Divaška jama is marked on the map with an unbroken *black line*, while the outline of its passages is marked with *grey*. The cave filled with sediments, which was detected with geoelectrical tomography, is marked with a *broken line*. Its continuation, i.e. an unroofed cave filled with sediments, which was detected in relief, is marked with a *dotted line*. The entrances to the caves are marked with *black crosses*, next to them, the Cave Registry number is written

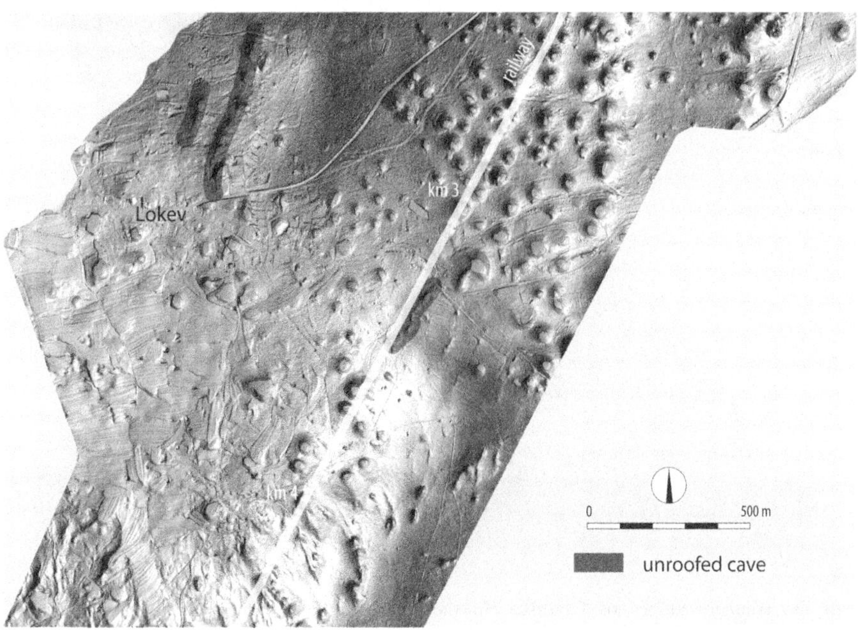

Fig. 13 DEM of Lipiški ravnik, Divaški kras and a part of Gradišče. Larger unroofed caves are marked with a *dark grey* colour

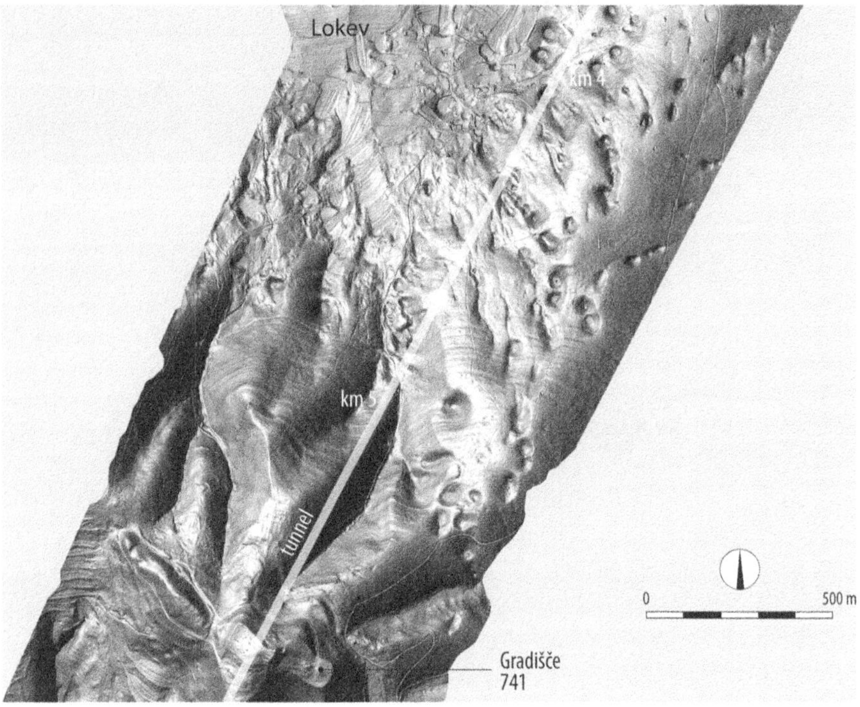

Fig. 14 DEM of Gradišče. The spike is located in the *bottom* part of the model on flysch rocks with a fluvial relief. The ridge running in the northeast direction is made up of limestone, which contains a few shallow dolines

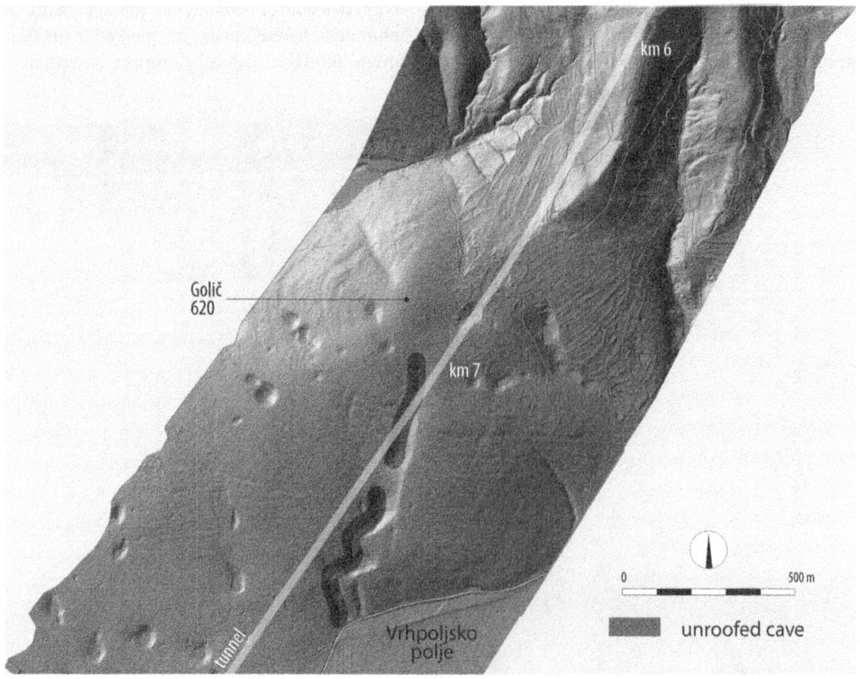

Fig. 15 DEM of the southern slopes of Gradišče. In the northern part they exhibit characteristics of a fluvial, ridge relief; in the southern part, dolines and a flat polje were formed on limestone near Vrhpolje. An unroofed cave or a more perforated limestone zone is indicated in the slope

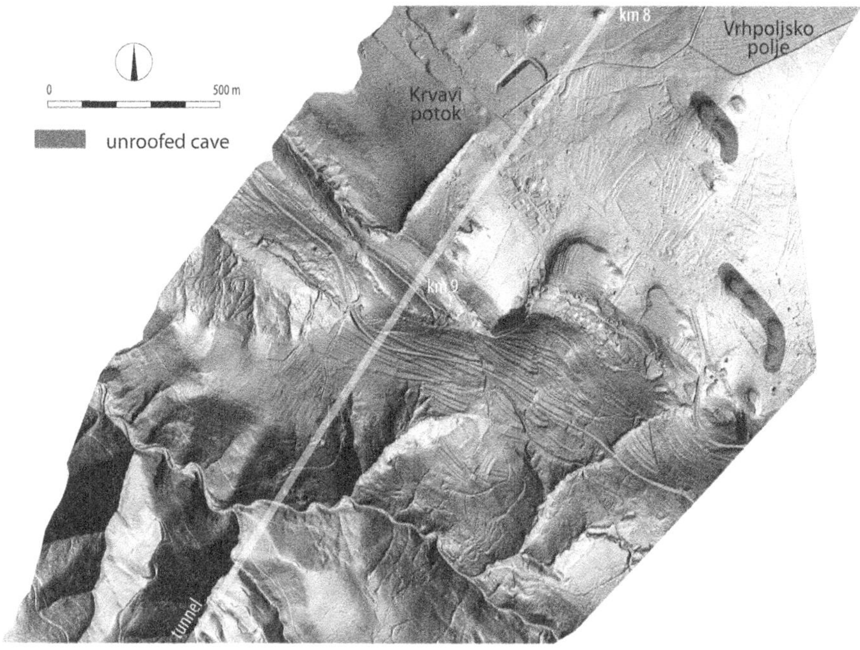

Fig. 16 DEM of the passage of Vrhpoljsko polje into the Glinščica valley. Two unroofed caves marked with *dark grey* colour are likely connected with the sinking of the brook that flows across Vrhpoljsko polje and largely also sinks there. Only during high water levels it flows across the now levelled riverbed on the surface into the Glinščica River

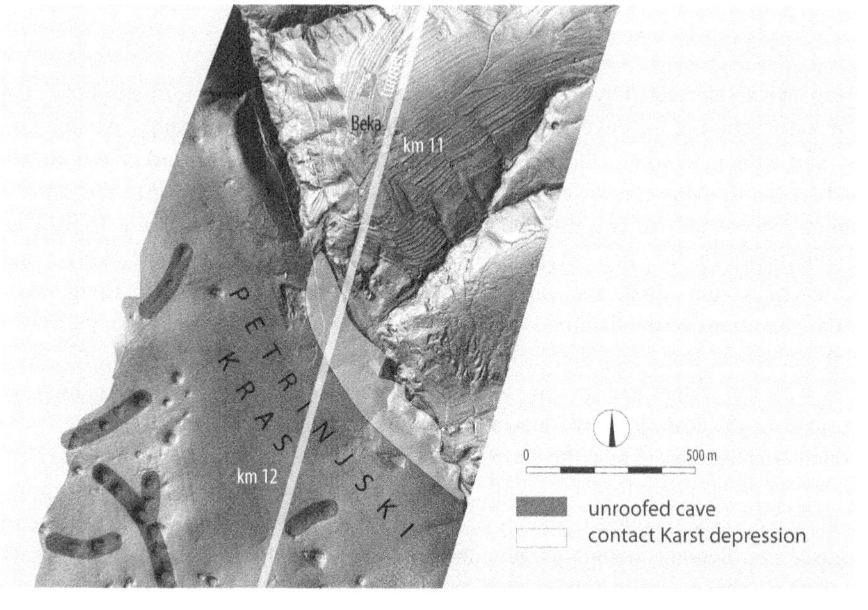

Fig. 17 DEM of the contact of the karst of Petrinjski kras and the flysch relief at the sinking streams of Ocizla. The unroofed caves are marked with a *dark grey* colour, while the hollow of the contact karst at the sinking streams of Ocizla is marked with *light grey* colour

Fig. 18 DEM of the karst area of Petrinjski kras, a part of the karst area of Podgorski kras. Unroofed caves, which are visible on the surface as elongated depressions or a series of depressions, filled up with clay and sands, are marked with a *dark grey* colour. A similar density of cavities can be expected at the height of the planned tunnel

Classical Karst, which is intersected by the route of the tunnels. With reconstruction of the development of the karst relief we can anticipate where the more perforated zones and types of cavities are to be expected.

The morphology shows at least two morphological stages of development. In the first morphological stage, still preserved in the relief, the surface was formed under the conditions of the dammed karst. On the west side, the Classical Karst was dammed by the flysch, which is why the waters had to flow towards the northwest. Due to the low gradient, the initially more diverse relief was flattened. Thus, the surfaces of the karst area of Divaški kras, Lipiški ravnik and Matarsko podolje were formed; higher segments of the relief, the Gradišče and Slavnik mountains, remained among them. Sinking streams flowed in from the Brkini hills. The inflow of rivers from the flysch caused great fluctuation of the groundwater, which was reflected in the formation of collapse dolines near the main drainage paths. At that time the area was probably drained towards the northwest, the Matarsko podolje lowland, towards the karst area of Divaški kras and afterwards to the Lipiški ravnik plain. The main drainage paths most likely avoided Gradišče.

At the karst area of Podgorski kras, which is structurally determined by narrow zones of flysch among limestone, the waters flowed along these contacts, which resulted in a high density of cave passages.

Later, a profound change occurred. It was probably triggered by tectonic movements, primarily the sinking of the Gulf of Trieste, which opened up the Classical Karst laterally towards the west. This enabled the drainage of the karst towards the rivers of Rižana and Osapska reka and the village of Boljunec. A high gradient was formed, which was used by the Glinščica

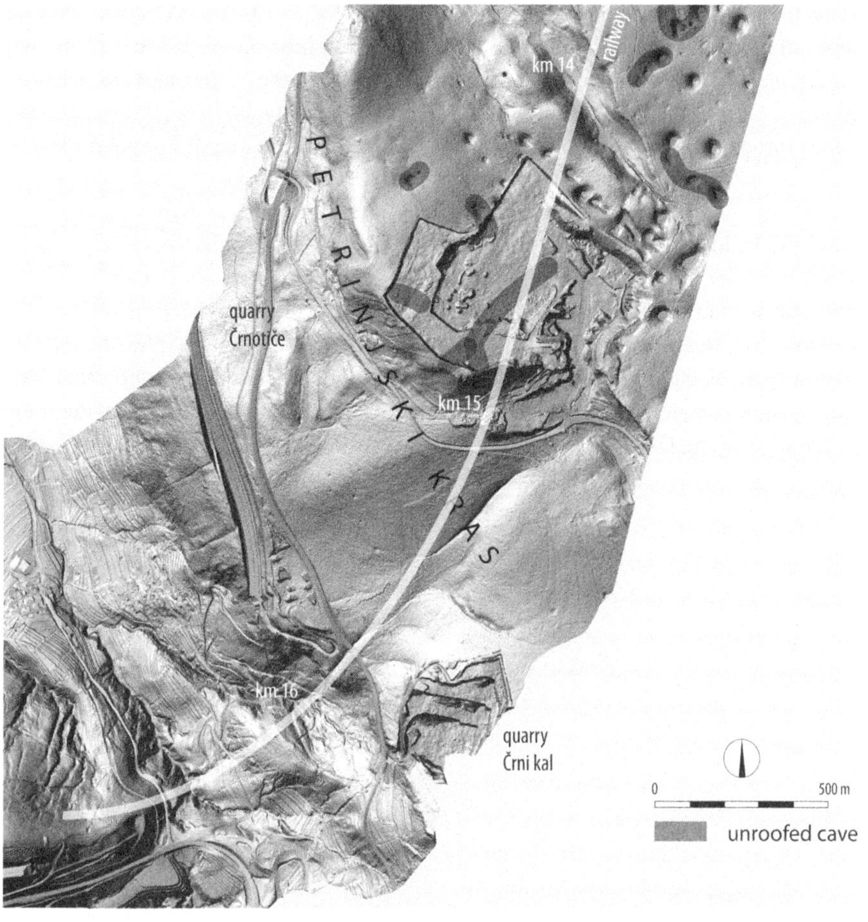

Fig. 19 DEM of the karst area of Petrinjski kras and of the Karst Edge near Črni Kal. This section contains the large Črnotiče quarry and south of it the Črni Kal quarry. Unroofed caves are marked with a *dark grey* colour. A similar density of cavities can be expected at the height of the planned tunnel

River on the surface, while in the underground the water level of karst water lowered, and hence the main channels descended lower down.

Under the present conditions, most of the relief is formed under the influence of precipitation. Throughout, the water table of karst water is located deep beneath the surface. It is similar in the karst area of Podgorski kras where deep, dry caves are accessible. Sinking rivers have no longer direct impact on the surface, except in proximity to ponors.

Current erosion is causing the oldest caves to open up to the surface. In exceptional cases the sections of their passages reach 20 × 20 m. Because the development of the deepening of the passages went on continuously and because these passages are also located in accessible, currently active caves, we can expect such passages to be spread throughout the karst massif.

4 Railway Construction on the Karst Surface

With a geomorphological analysis of the karst and non-karst surface above the planned route we cannot precisely determine where and of what kind the cavities exist in the underground. The morphological analysis is complemented with speleological and hydrological observations and enables the determination of zones and levels at which one can expect greater porosity or the location of important cavities.

The chief geomorphological findings that we must bear in mind are the following:

- The topography of the surface on the limestones indicates that today all the precipitation water from the area of the tunnels flows through the limestones karstically. This percolating water flows dispersedly to a great depth and does not create larger cavities. Creviced corrosion shafts are possible anywhere on the tunnel route.

- The route passes through larger morphological units that point to common developmental features and in which one can expect similar types of cavities.
- Larger cavities may be located throughout the entire limestone mass; however, greater density is probable beneath the surface of flatlands, in the karst area of Divaški kras, in the Lipiški ravnik plain (0.0–4.0 km) and in the karst area of Podgorski kras (11.5–16.0 km). There are not enough morphological signs for the area of the Gradišče mountain and the edge of the Matarsko podolje lowland near Vrhpolje (4.0–9.0 km) from which one could deduce the type and dimensions of the karst cavernosity.
- In the karst area of Podgorski kras or Petrinjski kras (11.5–16.0 km) the main courses on the surface of the unroofed caves run parallel to the contact with the flysch. Similar conditions can be expected at the depth at which the tunnel T2 will be dug.
- During the construction of the railway line on the surface, the route may encounter cavities, either empty or filled up with fine sediments, yet this cannot be predicted with the morphology of the surface.

Caves

1 Cavernosity of the Classical Karst

1.1 The Number, Size and Location of Caves in the Classical Karst

There are 777 known caves in the Classical Karst (Cave Registry; Fig. 1). Their number is growing, due to karstological and speleological research, which is also deepening the knowledge of the development of the karst. During activities that affect the karst surface, such as motorway construction, numerous old caves and shafts are being opened. New segments of already known caves are often found.

The longest caves are: Kačna jama (15,181 m), Škocjanske jame (5,800 m), Lipiška jama (1,400 m), Jama 1 v Kanjaducah (1,332 m), Vilenica (841 m), Brezno v Stršinkini dolini (800 m), Jama Sežanske Reke (699 m), Divaška jama (672 m), Škamprlova jama (585 m), Gustinčičeva jama v Blažčevi dolini (557 m), Križmančičeva jama (460 m), Velika Šprinčnica (400 m), Jama v Partu pri ogradi (396 m), Čebulčeva jama (379 m) and Brezno pri Risniku (300 m). The deepest caves are: Brezno v Stršinkini dolini (340 m), Jama Sežanske Reke (340 m), Jama 1 v Kanjaducah (329 m), Kačna jama (280 m), Škocjanske jame (250 m), Lipiška jama (250 m), Brezno pri Lipniku (230 m), Lipiško brezno (210 m), Vilenica (190 m), Mejame (173 m), SRT 1 (170 m), Jama na Konjičih (153 m) and Čebulčeva jama (143 m). The Labodnica cave located beyond the border, in Italy, is even deeper (319 m). Relatively short and shallow caves are more common. The average length of the known caves in the Classical Karst is 86 m, whereas their average depth is 30 m. The length of all the passages reaches 65 km. The entrances to the caves are situated between 31 and 650 m above sea level. The majority of the cave entrances (71 %) are found between 300 and 500 m above sea level.

The average density is 1–1.5 cave per km^2 of the karst surface area. Entrances are more densely distributed, particularly between the villages of Lipica, Orlek and Sežana, and in the vicinity of the village of Divača where the caves are the largest. Caves are most common in the flat karstified area north of Sežana, with as many as 22 caves per km^2, while in the broader surrounding area there are 9.8 caves per km^2 of the karst surface area. The average cavernosity rate is 572 m of caves per km^2 of the karst surface area. The caves are relatively spacious; the largest caves amount to from 10,000 to 200,000 m^3. With two million m^3, the Martelova dvorana hall at the end of the Škocjanske jame cave is the most spacious one in Slovenia.

The data, obtained during the construction of motorways, inform us that after construction and earthmoving works the knowledge of the density of caves is completely altered. In an area where there were no known caves, as many as five caves open up per kilometre on average.

The data were obtained from the known caves. They are, of course, not fully representative. During the motorway construction, which went several tens of metres deep into the surface, more than 350 cavities were discovered across 75 km of the route. Two-thirds of them were parts of the passages of old caves; many

M. Knez et al. (eds.), *The Beka-Ocizla Cave System*, Cave and Karst Systems of the World,
DOI: 10.1007/978-3-319-04456-9_3, © Springer International Publishing Switzerland 2015

Fig. 1 Caves in the Classical Karst (from the Cave Registry)

of them are already unroofed, and the rest are shafts. Prior to construction, there were only three known caves on the route. In larger funnel-shaped dolines a number of smaller fissure entrances to the shafts were opened up.

Surface karst phenomena, dolines and collapse dolines show distinct karstification. Larger collapse dolines may have cut off or buried the former cave passages, which are not accessible today.

1.2 Types of Caves and Their Shape

Caves can be divided into those through which streams of water are or were passing at different speeds and into shafts formed by percolating water. Old caves have preserved traces of the streams of water in the shape of the passages, in rock formations and sediments, whereas the flowstone has preserved traces of the percolating water. The Cave Registry records the following classification of caves: 5.3 % of horizontal caves, 26.5 % of caves with shafts and flats and oblique caves, 7.2 % of abris, 38 % of shafts, and 23 % of oblique and gradient shafts. The most common caves are the entrance shafts. Some of them are deep: the Lipiško brezno shaft is 208 m long and the entrance shaft to the Kačna jama cave 186 m (Fig. 2). On the walls of the shafts, through which water percolates from the permeable karst surface, channels or recesses have formed, while dry, old shafts are often coated or filled with flowstone. Old, dry caves are often split by shafts. The spaciousness and shape of the shafts are also affected by the fissures along which they were formed.

There are eight water caves. The largest Slovenian sink cave is Škocjanske jame; the Mejame cave is a periodic stream sink. They were formed by waters flowing from the flysch Brkini hills. The river that sinks in Škocjanske jame also appears in the caves Kačna jama, Jama 1 v Kanjaducah, Brezno v Stršinkini dolini, Jama Sežanske Reke, Labodnica, Jama Lazzaro Jerko, Jama Lindner and Golobje brezno. It emerges at the edge of the Gulf of Trieste in the Duino springs where waters from the Vipava valley, the Soča gravel and from the surface of the aquifer are combined. The water table of groundwater is also reached in the caves Drča jama, Jama Dolenca and Preserska jama. Occasionally, smaller streams flow from the field by the Mohorini homestead into the Mohorini sinkhole. In the Kraljeva jama cave there are also traces of a periodic, smaller stream of water.

1.3 Development of the Aquifer

With regard to the development of the through-flow and outflow aquifer, the known caves in the Classical Karst can be divided into those that streams of water flow through, i.e. sink caves and rare through-flow caves, into old and dry caves where there are traces of water flows and which are located higher up the aquifer; and into shafts through which water percolates from the permeable karst surface to the groundwater. Today, in the northern part of the railway route, its water table is up to 200 m deep and more; in certain spots in the southern part it is much higher. The various caves, criss-crossing the aquifer, reflect its development (Fig. 3).

Old caves with traces of water flows and flood periods, and with various sediments or flowstone are the remains of past, most distinct periods of the formation of the karst aquifer's underground. Works during the construction of motorways discovered the oldest caves. Due to the lowering of the karst surface, they have thin roofs or none at all. Two-thirds of the

Fig. 2 The layout of the Kačna jama cave (modified from Polák et al. 2012; Gams 2004)

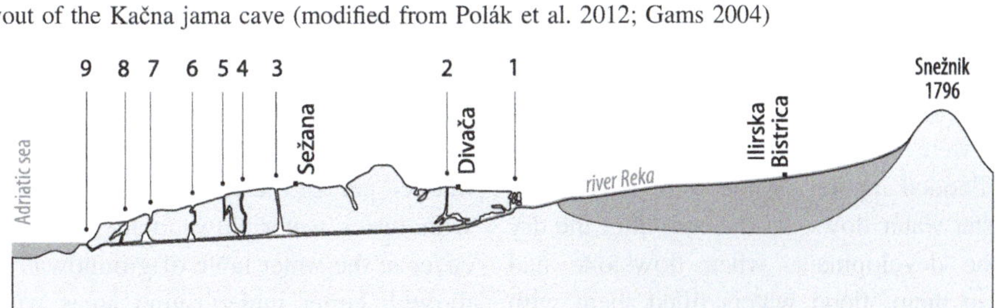

Fig. 3 The section of the karst between the Snežnik mountain, Škocjanske jame cave and the Timava springs (Gabrovšek and Peric 2006). The caves: *1* Škocjanske jame, *2* Kačna jama, *3* Jama 1 v Kanjaducah, *4* Brezno Sežanske Reke, *5* Jama v Stršinkini dolini, *6* Labodnica, *7* Jama Lazzari Jerko, *8* Jama AF Lindner, *9* Timava springs

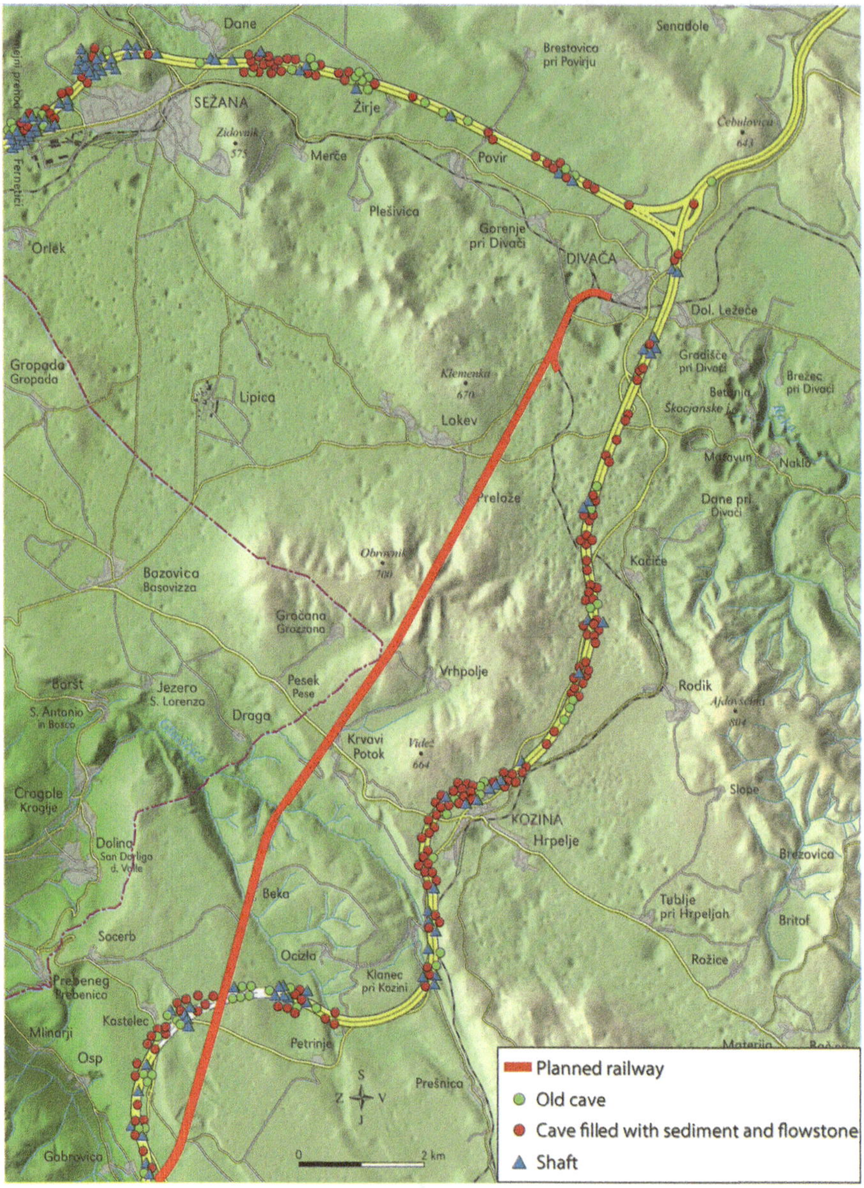

Fig. 4 Caves which opened up during the construction of the motorway between Divača, Fernetiči and Črni Kal, and the planned railway line (from Knez and Slabe 2007)

caves are filled with fine-grained sediment, and less with gravel and sand, or with flowstone or younger rubble. They were initially formed by slow water flows in the flooded aquifer. Some were later transformed by faster water flows. At the end, after the dry periods of the development, when flowstone had accumulated in them, flood waters filled them with fine-grained sediment (Fig. 4). In the study of the cave rock relief (Slabe 1995), a cross-section of the aquifer can reveal different periods of development and various factors of cave formation. In the old caves, traces of slower water flows, which formed passages in the flooded part of the bundle of passages, are intertwined with faster water flows, which are characteristic of caves at the water table of groundwater or which flow through larger underground areas with a free water level. When floods filled cavities with fine-grained sediment, they often caused the water to occasionally

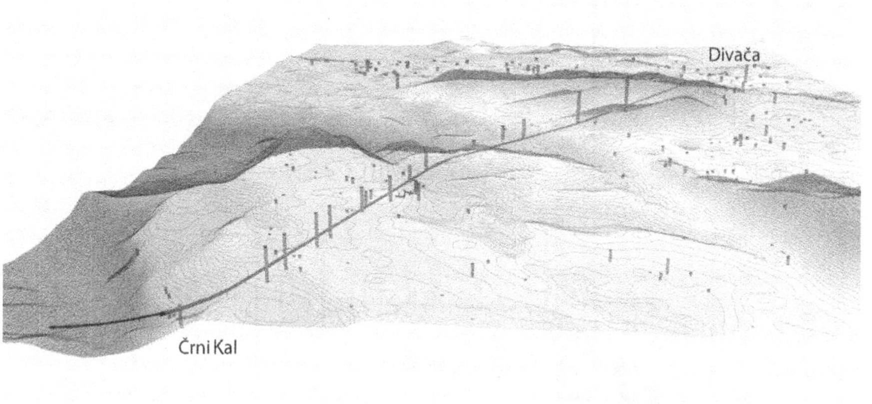

Fig. 5 The route of the railway line with delineated caves and boreholes. View through surface to the underground

flow above the sediment, reshaping the cave circumference.

When they were still surrounded by flysch, the karst aquifer limestones were closed in and the groundwater was dammed, which preserved the surface flows. These are said to have left traces on the karst surface in dry valleys and sediments, as established by karstologists. We are tracing slow karstification of the aquifer, which often occurs in leaps, with gradual lowering of the water table of groundwater, which is primarily linked to the vertical tectonic segmentation of karst regions and to the height of the lowering marginal flysch water barrier. The levels of distinct cavernosity, except for the present-day one and the one beneath the very peak of the Karst Edge, have not yet been discerned. Occasional minor fluctuations in the water table of groundwater are mainly the result of changing climatic conditions. In certain places, flysch areas remained for a longer period of time. Water flowed off these into the karst underground. This is also demonstrated by the fragile flysch pebbles in caves in the middle of the Classical Karst, far away from the today's flysch edge.

The water from the permeable karst surface is percolating dispersedly through shafts and fissures into the underground and partially reshaping old caves. Numerous shafts were discovered during the construction of motorways. Most of them were found at the bottom and sides of larger, funnel-shaped dolines. Most of the shafts were without a visible natural entrance. The narrow and impassable, fissured entrances were revealed after soil had been removed. The more spacious shafts were most often discovered and explored when larger cuts and tunnels were dug, i.e. several tens of metres beneath the surface. This is the result of the collection of the water that is percolating dispersedly through the permeable surface; the more spacious shafts —even those with a natural entrance, for the surface has been greatly lowered—are the collectors of this water.

1.4 Caves in the Area Between Divača and Črni Kal

There are 177 caves known in the broader area of the section of the future railway line between Črni Kal and Divača (Figs. 5, 6). Data on the caves are taken from the Cave Registry. The Beka-Ocizla cave system is discussed separately.

Most of the entrances to the caves, as many as 106 or 63 %, are located between 400 and 500 m above sea level. Only one cave is located below 100 m and there is likewise one cave between 200 and 300 m; there are 4 caves between 100 and 200 m, 27 caves above 500 m and 28 caves at an altitude between 300 and 400 m above sea level.

Over 70 % of the caves (118) are shafts, caves with shafts and flats or oblique caves, and oblique and gradient shafts (the nomenclature is in accordance with the Cave Registry). There are 19 horizontal caves (11 %) and 16 abris (10 %), two cave systems (Škocjanske jame and Kačna jama), while two caves are defined as periodic stream sinks and two caves as periodic springs.

Over half of the caves (57 %, 95 caves) are less than 20 m deep, 33 caves (20 %) are deeper than 50 m,

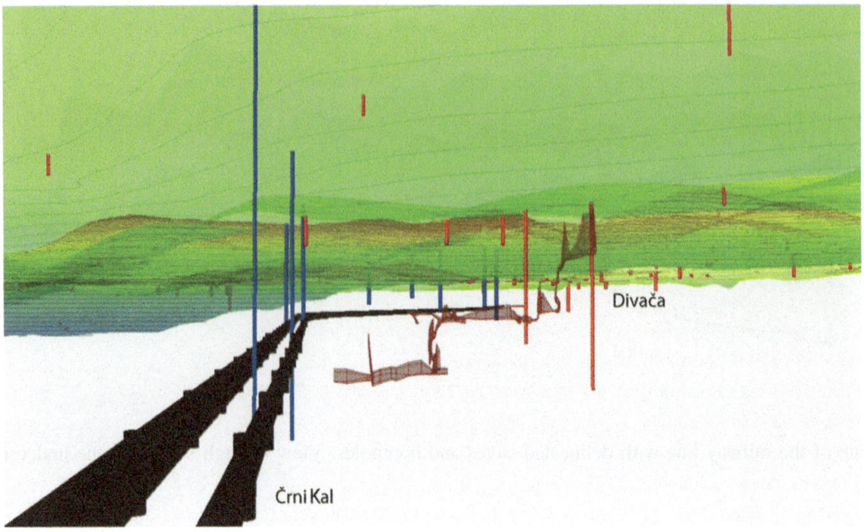

Fig. 6 The route of the railway line (*black*) with delineated caves (*red*) and boreholes (*blue*). View through the underground to the surface

and 15 caves (9 %) deeper than 100 m. Slightly less than a quarter of the caves have a depth between 20 and 50 m.

In the broader area, two caves are longer than 5,000 m and 4 caves longer than 1,000 m, which amounts to merely some 2 % of all the caves. Caves that are longer than 100 m stand for a poor fifth of all the caves (18 %, 30 caves), whereas over one third of the caves (36 %, 60 caves) are shorter than 20 m. Almost half of the caves are shorter than 100 m and longer than 20 m. When considering the length of the caves, we must bear in mind that the presented length of a cave is in fact a sum of the "horizontal and vertical lengths".

The most perforated part of the studied karst territory is the broader area of the Škocjanske jame cave, whereas in the southern part it is the broader area of the Beka-Ocizla cave system and the surroundings of the Kastelec road-tunnel (the Brezno na Škrklovici shaft), where the biggest cave opened up during the construction of motorways across the karst (Figs. 7, 8, 9). It is precisely in the area of the contact between the flysch and limestone, along which the Beka-Ocizla cave system has developed, that the chances of coming across caves of greater dimensions are very high.

Our research is beginning to show more and more frequently that large caves can also open up deeper beneath the surface as a vestige of the development of the karst aquifer.

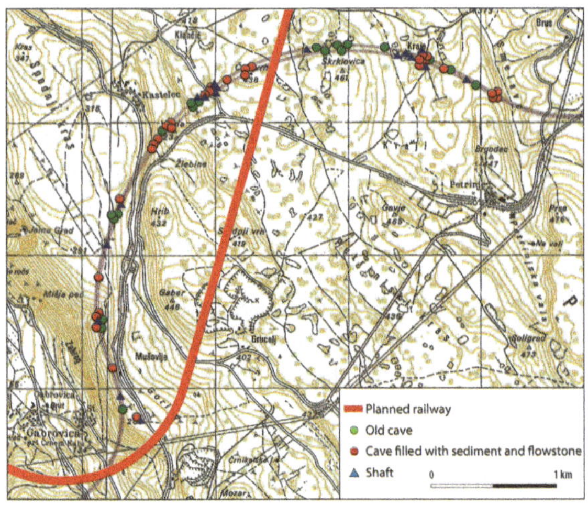

Fig. 7 Caves on the route of the motorway between Klanec and Črni Kal with the planned railway

Thus a slightly denser presence of caves can be found between the 1st and 3rd kilometre, 11th and 12th kilometre, and between 12.5 and the 13th kilometre of the railroad (see Fig. 21).

1.5 Caves at a Distance of 1 Km from the Planned Railway Line

In the area of the future Divača–Koper railway line we have speleologically and geologically inspected the

Fig. 8 A cave discovered while digging a motorway tunnel (*photo* F. Hrvatin)

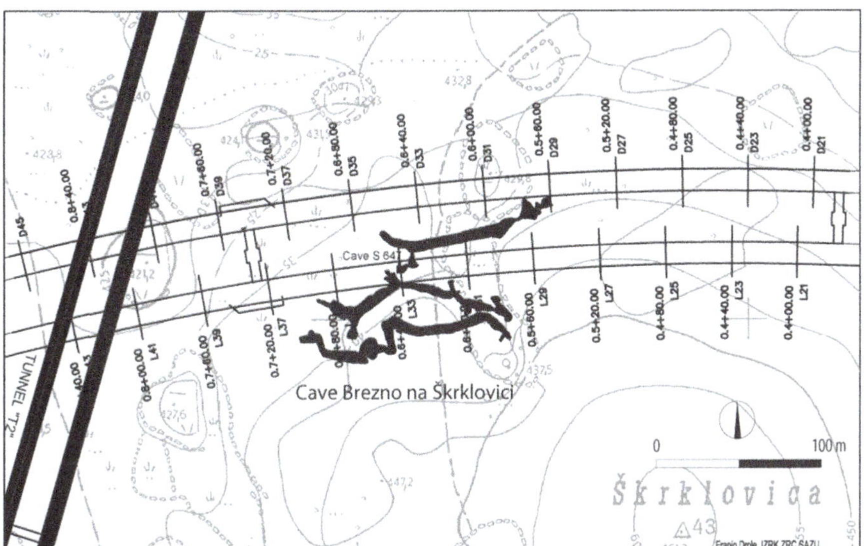

Fig. 9 A preserved cave network accessible from the Kastelec motorway tunnel with the planned railway

known caves located near the first two railway tunnels and registered in the Cave Registry (Figs. 10, 11; Table 1). We were interested in the location of the caves with regard to the proximity of the tunnel route, the depth and length of the caves and the possibility of the passages continuing into the vicinity of the tunnel route.

Fig. 10 The locations of caves, pinpointed anew in the field

We have inspected 39 cave entrances on the field and pinpointed their correct locations (Fig. 10). At the entrance we measured the basic geological elements such as directions and dips of the fissures, faults and bedding planes. Some of them are briefly presented in Table 1.

The cave Gabroviška jama is situated some 60 m higher than the overthrust of limestone on the Eocene flysch. Flysch sediment has been deposited on the bottom of this horizontal cave. The length of the cave is 21.5 m and its depth 3 m. The passage runs in the general E–W direction and is 350 m away from the planned railway route towards the west.

The Jama S 8 cave has 3 vertical entrances. They are geologically and structurally conditioned entrances and have developed along the fissures in the general N–S direction. The cave is 10 m deep and 30 m long. The Alveolina-Nummulites limestone is massive; a direction of the bed dip of 270/30 was measured north of the northernmost entrance. The cave is located 100 m away from the planned railway tunnel towards the west.

The Jama S 7 cave has developed along fissures, in the Dinaric direction (NW–SE). Alveolina-Nummulites limestone of a thickness around 1 m or less dips towards the southwest by 20° (250/20). The cave, which can be accessed through a 5 m deep shaft, afterwards descends steeply to a depth of 16 m. The length of the cave is 26 m. The main passage has the Dinaric NW–SE direction and it approaches the route of the tunnel to 225 m.

The Acijev spodmol abri has been described in an archaeological publication (Turk et al. 1992). It is an archaeological object of great importance, in which remnants from the early Holocene can be found. The abri is 300 m away from the future railway route.

The Brezno S 9 shaft is situated south of the Kozina–Koper main road on the Petrinjska gmajna common. The shaft opens up on the southwestern part of the doline that is situated right next to the main road. In order to enter the 20 m deep shaft, the entrance section first had to be dug out. It has developed in a fissure system with the dip direction of 240°. The Alveolina-Nummulites limestones dip at an angle of 20° towards the northeast. The shaft is located 225 m away from the railway route towards the east and will not intersect the planned tunnel.

The position of the Brezno na Škrklovici shaft was re-examined and it has been ascertained that the

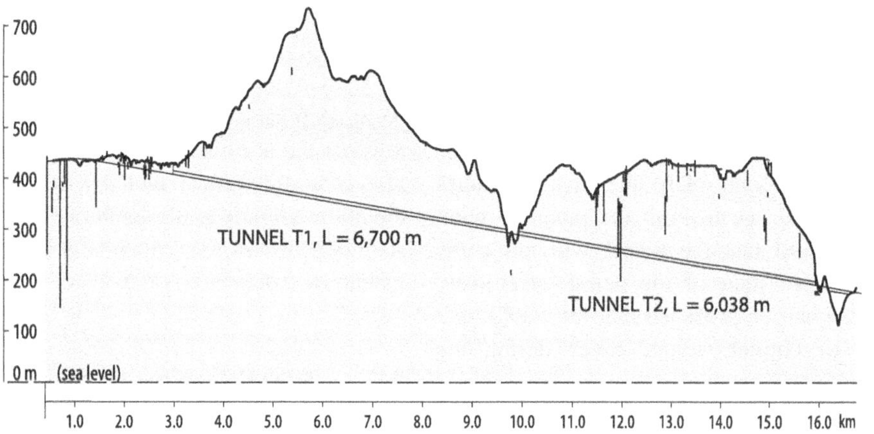

Fig. 11 A sketch of the caves (at a distance of 1 km from the planned railway line), shown in profile of dissected relief (see Fig. 10)

Table 1 The location of caves, pinpointed anew in the field, with the correct location of the entrances at a distance of 1 km from the planned railway line

Reg. No.	*x*	*y*	Distance between the cave entrance and the axis of the railway line (m)
238	413,190	48,440	11
636	413,982	50,712	26
723	414,122	50,500	154
729	414,170	50,450	211
741	418,630	59,530	235
955	419,085	60,275	349
1004	414,175	50,396	227
1005	413,370	50,053	450
1022	418,040	57,645	151
1390	413,734	49,487	821
1393	413,474	49,379	102
1578	412,740	46,360	280
1579	412,655	46,480	130
1597	412,560	46,600	698
1598	412,460	46,860	722
4528	413,220	47,590	160
4529	413,330	47,650	260
4815	419,195	59,310	370
5245	414,300	50,160	402
5247	413,601	50,735	404
5772	413,930	50,780	90
5940	418,520	59,170	165

(continued)

Table 1 (continued)

Reg. No.	*x*	*y*	Distance between the cave entrance and the axis of the railway line (m)
6167	415,155	52,880	181
6194	418,560	59,120	106
6960	413,040	48,030	168
6961	412,990	48,250	231
6963	415,820	53,460	387
7133	413,660	48,800	251
7591	417,730	57,545	72
7593	413,720	48,780	316
7643	413,075	47,450	70
7789	413,397	50,119	449
8527	413,509	48,384	228
8539	418,850	58,980	227
8540	418,830	59,017	191
8542	417,970	58,450	299
8543	417,960	58,440	303
8548	418,420	58,380	129
8909	413,860	50,130	14

location of the entrance was pinpointed in the Cave Registry. The entrance shaft with a depth of around 10 m has developed in fissures that have a Dinaric and cross-Dinaric (NE–SW) direction. The direction of the dip of the Alveolina-Nummulites limestone is 20° towards the southeast where the passage likewise descends to. The entrance to the cave measures 5–6 m in diameter. The length of the cave is 200 m and its depth 115 m. The cave is situated on a slope, some 225 m northwest from the Škrklovica hill (461.3 m above sea level). The entrance to the cave is some 150 m away from the planned railway tunnel towards the west. In the ground plan the cave continues towards the east, i.e., away from the planned route; hence, it is believed that the tunnel will not come across this cave. In light of the dense cavernosity of this section it can be predicted that there will be a great chance of coming across caves during the construction.

Two directions of fissures are predominant in the Udor na Škrklovici cave, namely the E–W and the Dinaric direction (NW–SE). The cave has 2 entrances. One entrance is from the collapse doline, while the other entrance is an 8 m deep shaft. The cave reaches 35 m in length and 10 m in depth. The thick-bedded Alveolina-Nummulites limestone dips by 10° towards the northwest. The direction of the passage is the Dinaric one. The cave is located around 75 m away from the railway route and does not extend as deep down as the planned route.

At Krvavi potok we conducted the geological and speleological exploration of two caves: Vh 1 and Čebina. In both caves the direction of the dip of the Alveolina-Nummulites limestone is towards the northwest by 25–50°. The entrance to the Čebina cave has been formed phreatically; the height of the narrow entrance (up to 1 m) is 1.6 m. Čebina is 31 m long and 12 m deep. The ground plan of the Vh 1 cave, which is 27 m long and 6 m deep, has a circular shape. Just below the narrow entrance, there is a slightly larger collapse hall.

The location of the Jazbina v Ravni cave was incorrectly marked on the 1:5000 topographic map, therefore its location was pinpointed anew. It has a smaller entrance (diameter = 0.5 m) and is shorter. The beds of grey micrite limestone dip towards the south by 15°. The thickness of the bed is 0.1–1 m. The entrance to the cave has developed at an intersection of two fissure zones, at 125/85 and 210/80. The prevailing direction of fissures is 210/80. The length of the cave is 7 m and its depth 4 m. The cave is located 100 m away from the railway route.

The entrance to the Brezno pri Trhlovci shaft measures 3–4 m and has developed in a strong crushed zone system at 100–110/80–90. The dip of the limestone is towards the southeast by 20°. Jazbina v Ravni and Brezno pri Trhlovci are classified under the karst area of Divaški kras.

The Golobivnica cave is located near Lokev. It is a 35 m deep karst shaft. The dip of the limestone bed is towards the north by 10°. The main geological structure is a sub-vertical fault by 10° in the northwest direction. Slightly to the south of Golobivnica the Lk 2 shaft was discovered, which is at least 17 m deep and its entrance diameter is 0.5–1 m.

2 Karstification of the Karst Discovered During Karstological Research of Boreholes in the Section Between Divača and Črni Kal

In the field planned for the construction of the railway tunnels we have karstologically inventoried 13 deep boreholes. This means that we have marked all the karst caves, flowstone, cave sediments, more fissured rock that provides good conditions for karstification, terra rossa that fills the fissures, etc. In the northern part of the tunnel T 1, the T 1-11 borehole was drilled entirely into Eocene flysch. Of the boreholes from the southern part of the tunnel T 2, the boreholes south of the Glinščica River, T 2-7 and T 2-8, were drilled entirely into flysch. The T 2-13, T 2-14 and T 2-15 boreholes pass from limestones into flysch.

A special attention was focused on the cores of 11 boreholes (T 1-7, T 1-8, T 1-9, T 1-10, T 2-9, T 2-10, T 2-11, T 2-12, T 2-13, T 2-14 and T 2-15) which have mainly been drilled into limestone.

In all of the 11 boreholes, which are from 150 to 350 m deep and have been drilled into carbonates, karstified areas have been determined (Fig. 12). In some of the boreholes, karstification was determined in only two spots, whereas in certain boreholes there were ten of such spots (T 1-8 and T 2-10); on average, there are 3 spots in each borehole.

According to data from the drillers, drilling mud leaked out from almost all of the boreholes at various depths in the karst. The outflows of the mud are clear signs of the openness of fissures and fault zones and, hence, of the important conductivity of the karst at various depths (from depths of a few metres through depths of around 100 m and all the way to a depth of over 240 m (T 2-11).

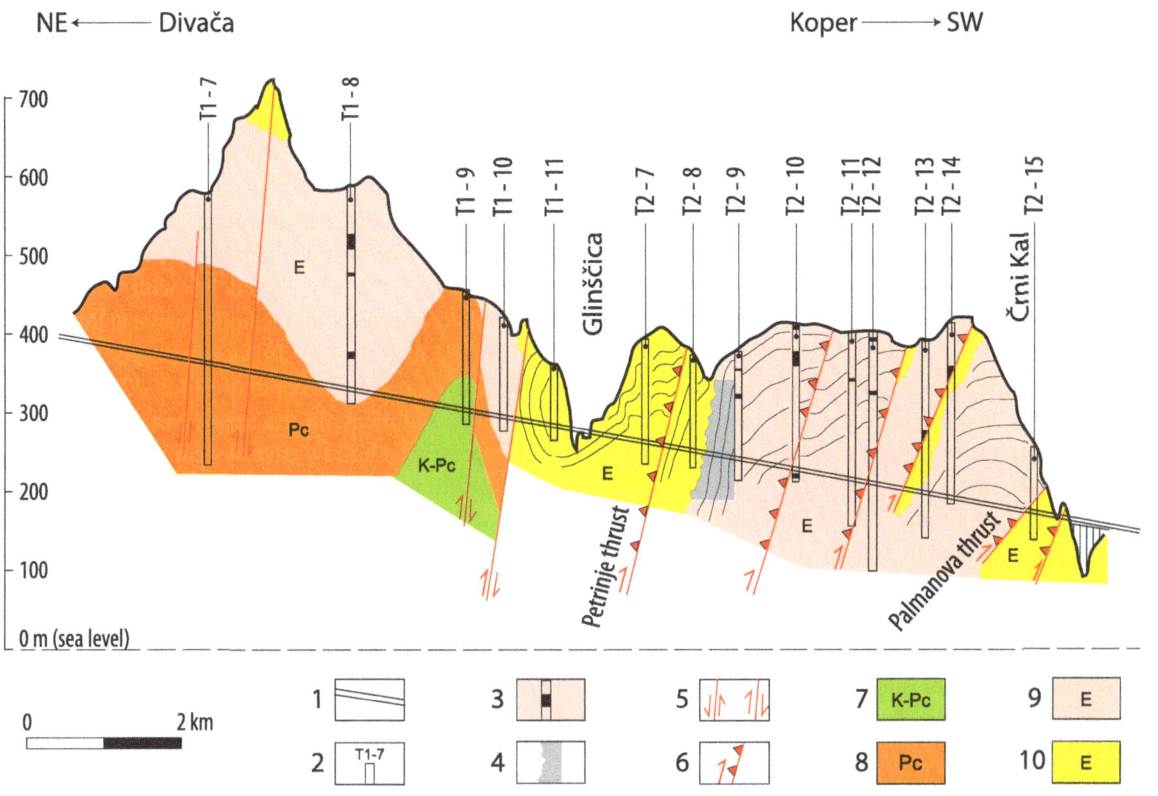

Fig. 12 Geological conditions on the route of the planned tunnel of the railway line between Divača and Črni Kal, boreholes with marked karst caves, flowstone and increased karstification. *1* route of the planned tunnel of the railway line, *2* position and designation of the borehole, *3* karst cavities, flowstone and increased karstification in the borehole, *4* position of the passages of the Beka-Ocizla cave system, *5* normal fault and reverse fault, *6* overthrust, *7* limestone of Liburnian formation (K-Pc), *8* Palaeocene limestones, *9* Alveolina-Nummulites limestones (Eocene), *10* Eocene rocks (in the *bottom part* there are marls, marly limestone and conglomerate, in the *upper part* there is flysch)

Larger (several metres in diameter) or smaller caves (karstified fissures, expanded to several decimetres) were discovered in the area of the northern part of the tunnel in one (out of four) borehole (T 1-8) and in the area of the southern part of the tunnel in five boreholes (out of seven). The largest cave with a diameter of 20 m (including cave sediments) was discovered in the T 2-10 borehole at a depth of 30 m; this is followed by a cave with a diameter of around 5 m at a depth of 60 m (T 1-8). The deepest cave was found at the depth of 192 m in the T 2-10 borehole. Due to the problems at the depth of 192 m where this cave was later encountered, the T 2-10 borehole was drilled three times with a few metres of distance between individual boreholes. In two attempts the aforementioned cave was encountered at the depth of 30 m, but not in the third attempt. This can aid us in assessing the volume or surface area of the karst cavity or its position in space.

In the T 2-13 and T 2-14 boreholes which partially pass from limestones into flysch, karst caves are located at the contact of carbonates and non-carbonates.

Based on the diameters of the cores, then on the 11 "spot surveys" in the boreholes over a distance of 10 km, on the determined karstification, on the discovered caves and on the fissures and zones in the karst that are open to water and through which the drilling mud flowed out several times in certain boreholes (T 2-10, T 2-11), we can deduce the incredibly high karstification level of the entire karst massif, through which the tunnel for the railway line will be drilled.

The T 2-8 and T 2-9 boreholes were located near the Beka-Ocizla cave system. None of them cut through the known passages of this cave system. Figure 12 clearly shows that the location of the known passages of the Beka-Ocizla cave system is in the vicinity of both boreholes. There is a considerable likelihood that during the construction of the railway tunnel in this area unknown passages will be encountered.

Other types of research were also conducted in the boreholes, such as geological mapping, inventorying of the geomechanical properties of rocks, tracer experiments, monitoring of the level and chemical properties of the water and recording in the boreholes and karst cavities.

The core from the T 1-7 borehole The T 1-7 borehole was drilled to a depth of 350 m (Fig. 12). The rocky cores are relatively uniform throughout the entire depth. The borehole did not hit any distinct karst phenomena, even though it is located in an area where such phenomena are otherwise very common.

What is primarily characteristic of the rocky cores, is the high level of fissuredness which is the result of tectonic events.

Traces of karstification were found only at the depths of 14.5–15 m and 38.5–39 m where loam is also present, which indicates that a cave could be located near the borehole.

Tectonic loam appears in several spots, e.g. at the depths of 59–61 m, 114.8–115.2 m and 130–130.7 m. In general, this rock is tectonically quite damaged and fissured, which has been ascribed to tectonic deformation.

During drilling, drilling mud leaked out at several depths (75–80 m, 82–83 m and 116 m), yet no traces of the presence of cave spaces were detected in the rocky cores.

The core from the T 1-8 borehole The T 1-8 borehole (Table 2), 275 m deep, is highly karstified. The upper part of the borehole is located at 596 m above sea level. The borehole encountered two levels of hollow, karst cave systems (Fig. 12). The first level is at the depths of 56.5–57.3 m and 59.4–59.7 m. It is most likely a single cave system at the depth of the borehole (56.5–59.7 m), which points to a karst cave of at least 3.2 m in height.

The next level of karst caves is at the depths of 60.4–65.5 m and 67.2–68 m, which could in fact be joined into a single level of 60.4–68 m. We are therefore dealing with a karst cave, 7.6 m in height, filled up with cave sediment and flowstone.

This is followed by the level of a cave filled up with sediments at the depth of 111.2–111.7 m; it is a cave, only 0.5 m high, which was hit by the T 1-8 borehole.

The third, deepest level of the cave filled up with sediments (loam) is at the depth of 214.2–215 m. This third level of the cave filled up with sediments is probably connected with the hollow cave at the depth of 217.2–220 m (Fig. 13). In fact it means that a cave system is located at the depth of 214.2–220 m, filled up with cave sediments in the upper part (214.2–215 m) and represents a hollow cave system in the bottom part (217.2–220 m). The height of this combined system is at least 5.8 m.

There is a gap of 157.5 m between both larger levels of the hollow caves.

The core from the T 1-9 borehole The core from the T 1-9 borehole is highly fissured and in some places broken virtually throughout its entire length. The loss of drilling mud occurred on three levels, which points to good conductivity of the rock into the surrounding rock block. The karstified areas were determined in only three spots at the depth of between 85 and 115 m; however, no karst caves were encountered.

The core from the T 1-10 borehole The T 1-10 borehole did not encounter any karst caves, for the ground consists of poorly karstified limestone. Terra rossa covers the fissures at least to the depth of 57 m. Limestone is somewhat more karstified at 91–100 m where fissures expanded by corrosion are present.

The cores from the T 1-11, T 2-7 and T 2-8 boreholes The T 1-11 borehole is located in the Eocene flysch (Fig. 14), which reacts well to HCl. It means that it contains a great deal of calcite, yet the rock is not karstified.

The T 2-7 borehole also drilled through Eocene flysch and did not encounter carbonate rocks.

The T 2-8 borehole is located near the Beka-Ocizla cave system and was drilled to the depth of 150 m. The borehole is located in Eocene flysch and did not break through to the limestones. Traces of karst phenomena were not detected in these rocky cores. In general, the flysch is highly crushed, especially at the depth of 10–15 m, around 40 m, and at the depths of 45–46, 49–55, 80.4–81 and 90.7–110 m. Calcite veins in the flysch appear at the depths of 6.8, 9.3, 10–18, 28.5–40 m and around the depth of 147 m.

Table 2 An example of the karstological description (the core of the T 1-8 borehole)

Depth (m)	Description of the core of the T 1-8 borehole
4–6	Fissures, vertical, filled up with a thin red-brown coating, karstified
6.2–28	Fissured, partially with terra rossa, rather compact, karstified
28–43	Compact limestone with rarer fissures at 2–5 dm
43.8–44.5	A vertical fissure
at 50	Breccia along the fissure (d = 1.5 dm)
56.5–57.3	A cave
58.2–59.4	Pieces of limestone (d = 0.1–0.5 dm) with flowstone
59.4–59.7	A cave
60.4–65.5	A cave with sediments, flowstone (d = 6 cm, located at 62.80 m), pieces of limestone, loam, crystallized loam, rubble, terra rossa binder, flowstone (d = 1 dm, located at 63.10 m)
65.5–67.2	Compact limestone with fissures, karstified
67.2–68.0	A cave with sediments
68.0–69.0	A vertical fissure, approximately 2 dm deep, karstified
69.0–77.6	Compact limestone, partially fissured
77.6–78.0	Vertical fissures
at 77.9	A 1 dm wide karstified fissure with sediment, goethite or limonite
78.0–78.1	A 1 dm wide cavity, its wall made of white, corroded limestone
78.1–88.0	Compact limestone
at 86	"Loss of eluting liquid"
88.0–88.8	A fissure with terra rossa, karstified
104.0–110.0	A vertical fissure
111.2–111.7	A cave filled up with sandy, dry sediment of a yellow colour, agglutinated in some places
112.3–112.7	Weathered limestone
at 118.2	Stylolite in compact limestone
122.0–122.8	Limestone, fissured, crystallized
123.0–130.0	Limestone, compact, with fissures
at 130.7	A fissure with terra rossa, karstified
at 137.5	A 5 dm long and 0.4 dm wide fissure with terra rossa, karstified
154.6–155.3	A fissure with calcite veins, wrapped, eroded by corrosion, open to a maximum of 0.04 dm
155.3–158.0	Compact limestone, partially fissured
158.0–169.7	Compact limestone with fissures
169.7–170.0	Shorter fissures with sediment
170.0–173.0	A vertical fissure filled up with red sediment, karstified
182.0–188.0	A vertical fissure filled up with red sediment, karstified
188.0–206.0	Vertical fissures
211.1–211.2	Limestone in pieces of up to 0.5 dm
214.2–215.0	A cave filled up with limestone gravel between loam
217.2–218.0	A cave filled up with washed out loam
218.0–220.0	A cave filled up with washed out loam
245.9–248.9	A vertical fissure, crushed, thin coating, tectonic (open stylolites)
253.5–256.5	A zone with fissures and vertical fissures
262.0–264.7	More compact limestone with fissures
269.0–275.0	Horizontal fissures

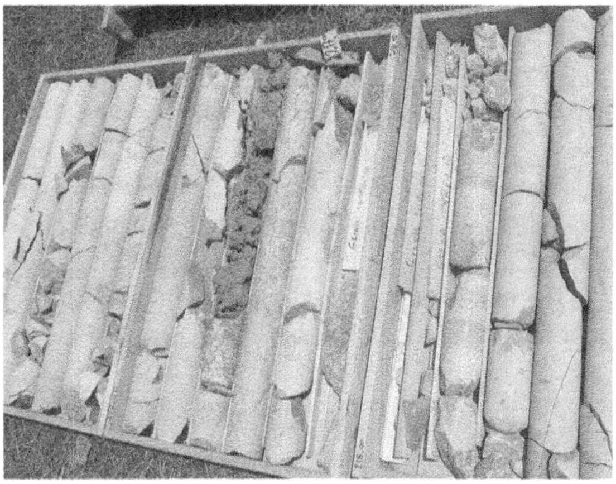

Fig. 13 A karst cave (an empty core of the T 1-8 borehole) at the depth of 217.2–220 m

The core from the T 2-9 borehole The T 2-9 borehole encountered a karst cave completely filled up with flowstone (Fig. 15) at the depth of 68.1–68.7 m; the drilling mud disappeared at this very depth (68–69 m).

The core from the T 2-10 borehole The T 2-10 borehole (Fig. 16) is located 429.3 m above sea level and is 200 m deep. A karst cave is located at the depth of 9–9.5 m. Flowstone is present at the depth of 30–31.7 m. A karst cave was encountered at the depth of 32–48 m, and cave loam that fills up the cave at the depth of 48–48.7 m. Cave loam was found also at 49–52.2 m. A cave system is probably located at the depth of 30–52.2 m and partially filled up with cave loam and flowstone or it is hollow. Thus the depth of this cave system is at least 22.20 m, which indicates greater karstification and a larger cave system.

The rock again shows some signs of karstification between the depths of 87.5 and 93.5 m, where red loam is present. Slight karstification also appears between the depths of 125–130.6 and 176.5–176.7 m. In some places the cores are tectonically highly fissured, especially at the depths of 75–79 and 82–84 m. Tectonic clay was also traced at the depth of 81.4–81.5 m.

Problems with drilling occurred at the depth of around 180 m. That is why another borehole was made a few metres away from it, which likewise encountered problems at the same depth. The third borehole made a few metres away from the first two progressed up to 200 m without greater difficulty. At the depth of 122.5 m a loss of drilling mud was detected, which indicates good karst conductivity.

A larger karst cave was encountered at the depth of 192–200 m.

The core from the T 2-11 borehole The T 2-11 borehole was drilled up to the depth of 250 m; its upper part is at 430.7 m above sea level. During drilling, the drilling mud was often lost, namely at the depths of 41–42, 58–59.5, 76.5, 240.5 and at 245.6 m.

The limestone is quite compact; lithology varies so that in certain places limestone is darker and richer in organic remains.

Fig. 15 Flowstone is located at the depth between 68.1 and 68.7 m and fills up the karst cave in the T 2-9 borehole

At the depth of 240.5 m where the drilling mud was lost, a karst cavity and fissures filled up with thin, yellowish loam were discovered. Terra rossa covered the fissures up to the depth of 64 m.

The core from the T 2-12 borehole The core from the T 2-12 borehole points to high karstification up to the depth of around 70 m; deeper down, karstification

Fig. 14 Lithological changes in flysch as seen in the core of the T 1-11 borehole

Fig. 16 The surface landscape and the T 2-10 borehole

Fig. 17 A part of the core from the T 2-13 borehole that shows a tectonic overthrust transition from limestone (*lighter cores*) to flysch (*darker cores*)

is smaller, despite the highly broken rock. What deserves mention, is the flowstone at the depths of 13–14.20 and 19.40–19.60 m, and a smaller karst cave at the depth of 65 m where the loss of drilling mud also occurred.

The core from the T 2-13 borehole The T 2-13 borehole reached the depth of 250.5 m; the upper part of the borehole is located 412 m above sea level. The borehole contains limestone up to 141.4 m; afterwards, it passes tectonically into flysch (Fig. 17) and then back again into limestone at the depth between 183 and 184 m.

A karst cave, filled up with red-brown solid clay and pieces of limestone with diameters up to 4 mm, is located at the depth from 136.9 to 137.1 m and only 4.3 m above the overthrust of limestone on flysch.

The core from the T 2-14 borehole The core of the T 2-14 borehole shows great fissuredness and brokenness of the rock. Slight karstification can be traced throughout. As expected, a karst cave is located on the contact of limestone with flysch.

The core from the T 2-15 borehole The T 2-15 borehole reaches to the depth of 130 m and passes from the limestone in the upper part to the bottom flysch. The upper part of the borehole is located at 257.5 m above sea level. The tectonic transition to the bottom flysch occurs at the depth of 82.2 m. Up to the depth of 43 m the limestone is highly crushed and in some places karstified. The flysch contains many calcite veins, some of them have been tectonically displaced.

3 Karstological Research in the Construction of Motorways Across the Slovenian Karst

One of the biggest projects, currently being finalized in Slovenia, is connecting the state with modern motorways. Almost half of Slovenia consists of karst and more than half of the waters from which we are supplied come from karst aquifers. Slovenia is the land of the Classical Karst, which gave the name to this unique landscape, developed on carbonate rocks, in numerous world languages, and in which karstology began to develop. Thus the delicate karst landscape demands that we are well familiar with it and that we endeavour to preserve it. It is an important part of our natural and cultural heritage.

For many years now, karstologists have been participating in the planning and construction of motorways in the Classical Karst (Kogovšek 1993, 1995a, b; Knez et al. 1994, 2003, 2004a, b, 2008; Knez and Šebela 1994; Šebela and Mihevc 1995; Slabe 1996, 1997a, b, 1998; Mihevc and Zupan Hajna 1996; Mihevc 1996, 1999, 2001; Kogovšek et al. 1997; Mihevc et al. 1998; Šebela et al. 1999; Bosák et al. 2000; Knez and Slabe 1999, 2000, 2001, 2002, 2004a, b, 2005, 2006a, b, 2007, 2009, 2010, 2011, 2012a, b, c, d). When choosing the route of the motorways and railway lines, we must first take into account the integrity of the karst landscape and the recommendations for avoiding the more important surface karst phenomena (dolines, poljes, collapse dolines, karst walls) and the already known caves. Special attention is focused on the impact of the construction and use of motorways on karst waters. Motorways are said to be impermeable; i.e., waters from the roadway are first collected in oil separators and then released into the karst in a purified state.

We have studied the impact of traffic roads on karst water. Kogovšek (1993, 1995b) has determined the composition of the contamination of the waters that flow from the motorways on a daily basis. The stagnant waters, smaller quantities of which were found in the caves by the traffic roads, contained traces of mineral oils (Knez et al. 1994).

We carry out karstological supervision during the motorway construction. We study the newly-discovered karst phenomena as an important part of our natural heritage; we advise methods for preserving

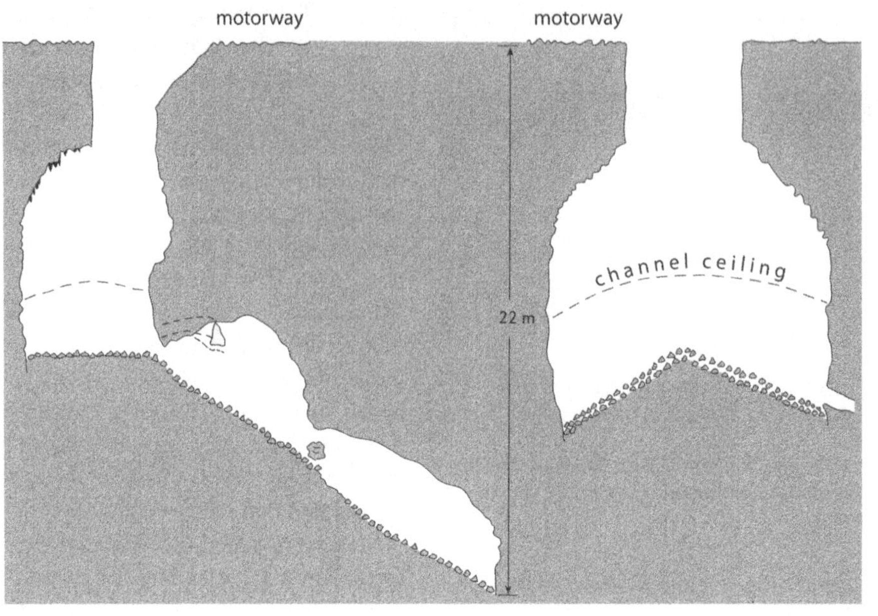

Fig. 18 Exploration of the cave whose roof collapsed due to the blasting during the motorway construction near Divača

them if the construction works allow for it; our new findings often aid builders (Fig. 18). We have made many new findings on the formation and development of the karst surface, the epikarst, and the cavernosity of the aquifer.

3.1 Karstological Supervision During the Construction

The removal of soil and vegetation from the karst surface and the earthmoving works on a larger scale during the digging of roadcuts and tunnels, revealed surface, epikarst and underground karst phenomena. Our mission is to study these phenomena as a part of natural heritage, to advise methods for their preservation and, of course, acquaint the builders with new findings. These findings aid them in overcoming obstacles during the construction.

The karst surface is dissected by dolines and unroofed caves (Fig. 19). Today, dolines reflect the way the surface is being shaped by atmospheric water, which is percolating through it vertically and then through the vadose part of the aquifer down to the groundwater. Dolines are more or less filled up with soil. On their bottom, shafts and crevices open up, through which water flows away. The soil must be removed from the dolines and the floors reinforced with rocks, stacked into arches, since the mouths of shafts are often smaller than the nearby cavities beneath them, and then the dolines must be filled up with layers of rubble. The unroofed caves are of similar shapes or more elongated. These are old caves that have "peeked out", due to the lowering of the karst surface, i.e., they no longer have the upper parts of the circumference. The fine-grained fillings have to be removed from them as well, in this case the old cave sediments, and then the caves must be filled up with rocks and rubble. Water could eventually remove these sediments and silt could appear on the surface.

The epikarst is criss-crossed with karren; they are especially visible in the Cretaceous limestone and less in the Palaeogene one; more crevices were opened up at the bottom and slopes of dolines. They are mostly covered with soil and their walls are dissected with subsoil rock forms. Due to the lowering of the karst surface, many shafts are already located directly beneath it.

Across the 75 km of the route of motorways, which had been built in the Classical Karst in recent years, more than 350 caves were opened up (Figs. 4, 7). With regard to the development of the aquifer, the caves can be divided into old caves through which water flowed when the karst aquifer was surrounded and covered

Fig. 19 A unroofed cave near Povir, sediments and flowstone are removed from it

with flysch higher up, and into shafts through which water is percolating vertically from the permeable karst surface to the groundwater. The deepest shaft measured 110 m. The old caves are empty or filled up with sediment; the latter form almost two thirds of all the caves, while one third of the caves are already unroofed.

The caves are being opened up during the removal of vegetation and soil from the surface; an especially large number of them were opened up during the digging of roadcuts. When the rock was blasted, their roofs caved in, while the cross-sections of the passages were preserved in the banks. The majority of the shafts were opened up on the bottom of dolines, after the soil and sediments had been removed from them.

All the caves have been explored, their plans drawn and their shapes and rock reliefs defined; we have taken samples of the sediments for palaeomagnetic and pollen research and flowstone samples for mineralogical research and dating. Based on the shapes of the caves and the geological features, we have predicted their continuations, which will be of particular use to the builders.

We have tried to preserve as many caves as possible. The shafts were the easiest to preserve. Their smaller entrances were sealed with concrete slabs. It was likewise possible to preserve old caves that had a firm circumference. The caves that were opened up because of the blasting and were located in crushed rock had to be blasted and filled up. The caves that had been intersected by the cuts and whose entrances lay on their banks were sealed with rock walls. Their circumference was too crushed, which made the caves unsuitable for further inspections, whereas in the case of caves that were filled with sediments, water could wash loam onto the roadway. We left one well-preserved cave open to be viewed by passengers crossing the border with Italy. The most interesting and well-preserved caves were fully protected and even though they are located beneath the motorway, they are accessible. Concrete pipes lead to them, which end in a closed shaft by the road. The largest cave system in the Kastelec road-tunnel is almost entirely preserved (Figs. 8, 9); only a small part was damaged by rock blasting when the tunnel was dug. The cave consists of three major passages. Below traffic belt they are connected with large concrete pipes (Fig. 20). In the side of the tunnel there is a special door with a smaller niche behind. At the bottom of the niche the lid covers the shaft.

We have also studied the consequences of different blasting activities inside the caves, which will aid in further construction and preservation of the karst phenomena.

3.2 New Findings About Development of the Karst Obtained During the Construction of Motorways

Unroofed caves are a special and common karst form. This surface karst form, which is still important today, is an already known phenomenon, yet it has not been comprehensively studied. It has not been given enough attention, seeing that the portion of such surface phenomena is much greater than it was originally presumed. The number of publications about unroofed caves is connected with the construction of new sections of motorways. The shape of a unroofed cave is a result of the type and shape of the cave and of the development of the karst aquifer and its surface under various geological, geomorphological, climatic and

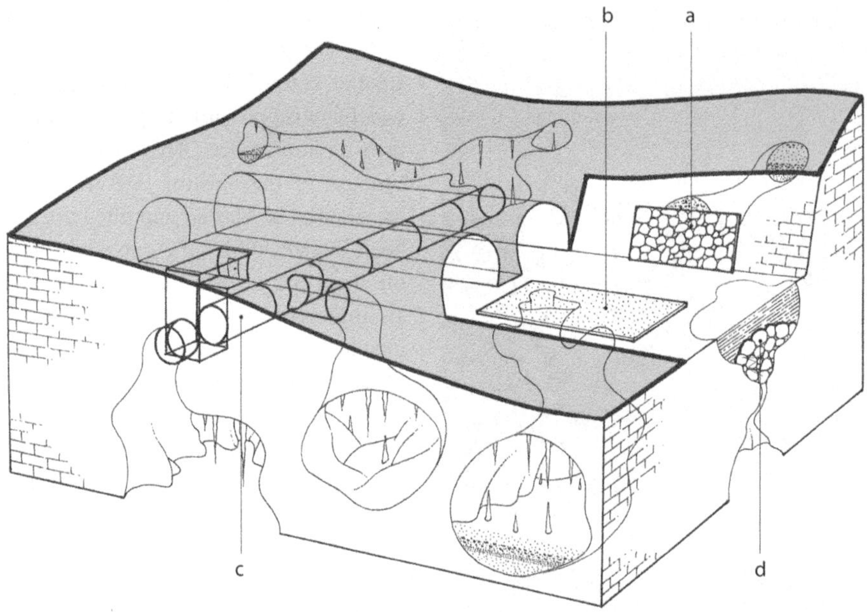

Fig. 20 Closing and protecting the caves. *a* in road cuts the caves are hidden behind rocky scarps; *b* the caves lying below the road are covered by concrete lids; *c* an entrance into the cave; *d* the bottom of karst fissures and the top of the shafts are often closed by arches of the rocks

hydrological conditions. The distinctiveness of the surface shape of a unroofed cave is dictated by the speed with which the sediments were removed from the cave in comparison with the lowering of the surrounding surface. On the surface we can discern soil and vegetation or zones of sediments and flowstone if the speed was slow; if it was faster, the unroofed caves on the karst surface resemble dolines or series of dolines or are elongated indentations. They are often a network of various old forms (caves) and present-day formation of the karst with dolines and shafts.

A great portion of the caves was filled up with sediments. The latter are most often flooded fine-grained flysch sediments, with beds of gravel in between. We have also taken samples of sediments for palaeomagnetic research; it has been ascertained that the sediments in the caves near Kozina and Divača are from the older Oldowan era. It has therefore been concluded that the caves were filled with sediment after the Messinian crisis when the Mediterranean basin was refilled with water, i.e. approximately 5.2 million years ago (Bosák et al. 2000).

All in all, the unroofed caves are becoming an increasingly distinct phenomenon on the karst surface, an important part of the epikarst and an outstanding remnant of the development of the karst aquifer. Our findings are also beneficial when planning various activities that affect the karst.

By dating the sediments we are identifying the oldest periods of the karstification of the Classical Karst and ascertaining that the oldest caves in the Classical Karst are much older than our karstological predecessors assumed.

4 Cavernosity Along the Planned Railway Line

In the following, we are presenting individual sections on the planned tunnels' T1 and T 2 route with types and number of caves we can expect in them.

Section A This section (Fig. 21) runs across Cretaceous limestones and limestones of the Liburnian formation. The Cave Registry records 11 caves in a two-kilometre zone along the route of the railway line, whereas 15 caves were discovered nearby in the same rocks during the construction of motorways.

It is expected that from 5 to 10 caves/km will open up in this section. Passages (old, dry ones created by the former discharge of the flows of water in the predominantly flooded zone of which Divaška jama cave is an example) might have greater diameters

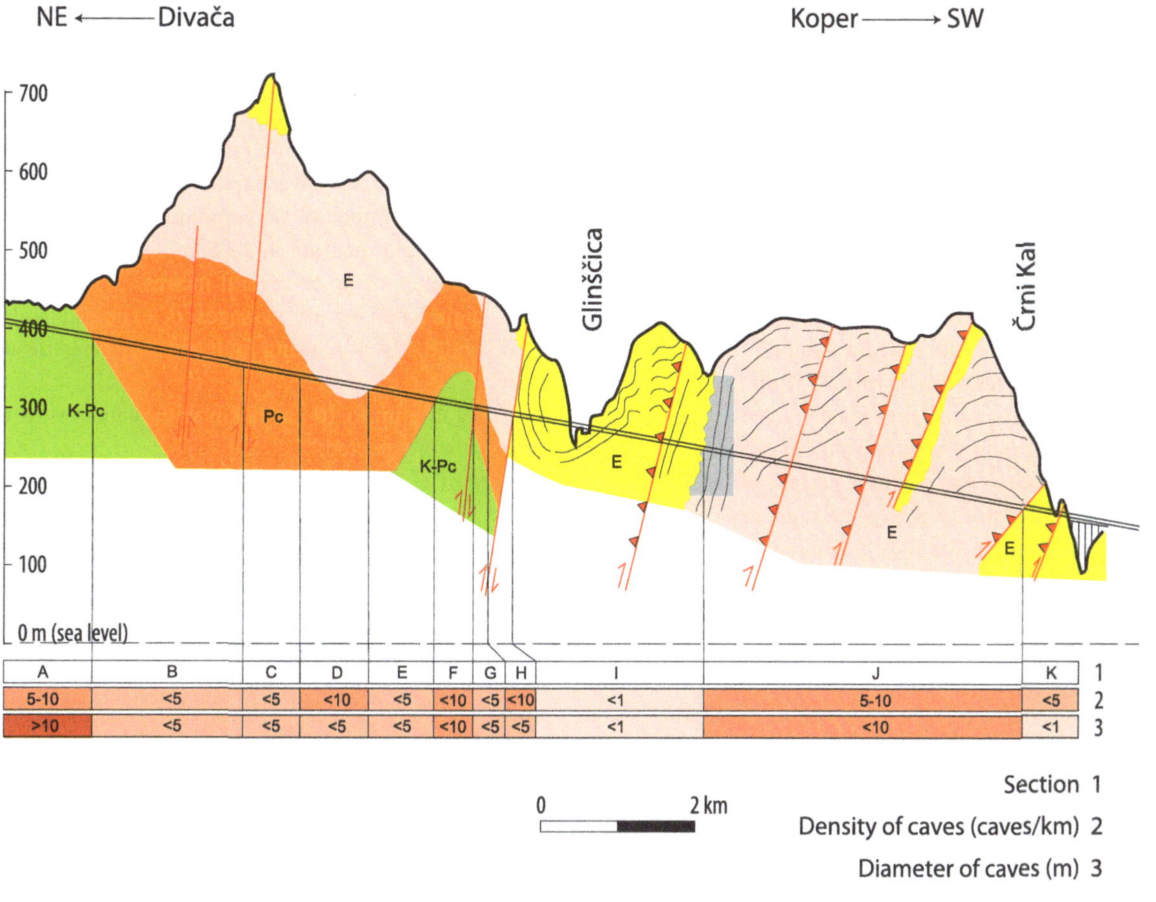

Fig. 21 Estimated density of caves and their average expected diameter

(over 10 m) and larger predominantly horizontal epiphreatic passages might also appear. It is anticipated that 50–66 % of passages will be filled up with fine-grained sediments and rubble (just below the surface), and deeper down also with flowstone. In between, (sub-)recent vadose shafts might appear whose diameters will measure up to 5 m. Their cross-sections will be round or creviced along distinct fissures. The caves will be located above the regional level of the karst groundwater and in some places we can expect more powerful jets and a hanging water body.

Greater cavernosity is expected at the Cretaceous–Palaeogene boundary at the transition of section A into section B.

From 5 to 10 caves/km are expected, some might have a diameter of over 10 m.

Section B This section passes through Palaeocene limestone. The knowledge of lithology and the borehole data show a smaller frequency of caves in the limestones of the Danian and Thanetian.

The Cave Registry does not record any caves in a two-kilometre zone along the route; however, several smaller caves were discovered in the motorway section, which were mostly filled up with sediment.

It is anticipated that the majority of the passages will measure up to a few metres in diameter and will be filled up with fine-grained sediment. We also expect vadose shafts with a diameter of up to a few metres and a depth of a few tens of metres. The concentration of vertical shafts will drop towards the end of the section, as vadose speleogenesis is becoming increasingly shorter, due to the synclinal geological structure.

The possibility of a hanging water body is small, yet greater than in the previous section.

Greater cavernosity is expected in the crushed area at km 4.840.

Up to 5 caves/km with diameters of up to 5 m are expected.

Section C This section passes through Palaeocene limestone that is covered with impermeable Eocene marl and flysch. Since the carbonate rocks are covered with flysch, vadose shafts are not expected. This area has a higher possibility of a hanging water body. We expect passages with diameters of up to a few metres, exceptionally up to 10 m if they are the less expected remains of old, now fossilized caves.

In light of the shorter period of speleogenesis, smaller cavernosity is expected; greater cavernosity is possible only in the locally limited inflows of surface waters from the flysch cover.

Up to 5 caves/km with diameters of up to 5 m are expected.

Section D As regards geological structure, section D passes through Eocene limestone that is not covered with impermeable rocks on the surface. We expect caves that were created in the phreatic or epiphreatic hydrological zones whose passages are (sub-)horizontal, and vadose shafts. The density of the shafts should increase from the beginning of the section towards its end, due to the length of speleogenesis.

Considering the high cavernosity of such limestone in the Karst Edge, we can expect greater cavernosity of this area, as indicated by the status of the recorded caves near this section. Due to the numerous activities affecting the Karst Edge (quarry, motorway construction, etc.), it has been studied much more thoroughly.

Up to 10 caves/km with diameters of up to 5 m are expected.

Section E This section passes through Palaeocene limestone. Lithological data and data obtained during blasting point to a smaller frequency of caves in Danian and Thanetian. The Cave Registry does not record any caves in this section; however, several caves were discovered in the motorway section which were partially filled up with sediment.

It is expected that the majority of the passages will measure several metres in diameter. Some of the caves might be filled up with sediments. Vadose shafts with diameters of up to several metres are expected.

In light of the geological profile, greater cavernosity is expected along the fault zones at km 7.1 and 7.3.

Up to 5 caves/km with diameters of up to 5 m are expected.

Section F The short section F passes through Cretaceous limestone of the Liburnian formation. The Cave Registry records one cave directly next to the route. In this section we expect segments of larger (sub-)horizontal phreatic and epiphreatic caves, whose diameters might measure up to 5 m, and vadose shafts with diameters of up to several metres.

A higher density of passages is expected at km 8.300 and 8.440, for strong fault zones are manifested there.

Generally, up to 10 caves/km with diameters of up to 10 m are expected.

Section G This section passes through Palaeocene limestone. The Cave Registry does not record any caves in this section; however, several caves mostly filled up with sediment were discovered in the motorway section.

It is expected that the majority of the passages will measure several metres in diameter. Some of the caves might be filled up with sediments. We expect vadose shafts with diameters of up to several metres and upright, narrow and long crevice caves created along distinct faults and fissures in the NW–SE direction. They can be up to 2 m wide and 10 or more metres long.

Greater cavernosity is expected at km 8.5 where the route crosses stronger fault structures.

Up to 5 caves/km with diameters of up to 5 m are expected.

Section H This section passes through Eocene limestone. The Cave Registry records one cave along the route. Vadose shafts with several metres in diameter are expected and so is a greater cavernosity at the contact of the Alveolina-Nummulites limestones and transitional beds and flysch at km 8.930.

Generally, up to 10 caves/km with diameters of up to 5 m are expected.

Section I This section passes through impermeable Eocene flysch rocks. Despite the carbonate binder in flysch rocks, which is in some places substantial, we do not expect greater cavernosity. There is a possibility of conduit channels at km 10.830, since lateral phreatic flows of water, independent of the locally impermeable flysch cover, may occur along the folds that are inclined longitudinally.

Generally, up to one cave/km with a diameter of up to 1 m is expected.

Section J This section passes mostly through limestones of Eocene age with rare, yet important hydrologically partially blocking wedges of marly and flysch rocks. In this section the Cave Registry records 20 caves, with a greater concentration between km 11.5 and 12.3 and between km 13.5 and 14.0.

Epiphreatic caves (traces of the former water levels) are found at various levels; the passages have diameters of up to 5 m and are mostly (sub-)horizontal. The area is being hollowed out also by vadose shafts with diameters of up to 5 m. This and the numerous larger caves formed along crevices were revealed by karstological research during the motorway construction. The research conducted at the Črni Kal quarry points to a 3.9 % cavernosity of the upper 19 m of Eocene limestone.

The limestone is very well, yet selectively karstified. Karstification is present in areas with concentrated active and fossilized flows of water at the contact with impermeable rocks.

Larger caves are found at the outflow of water from the Eocene flysch in the area of the Beka-Ocizla cave system between km 11.280 and 11.900.

Greater cavernosity is also expected along the fault zones between km 12.430 and 12.530. Great cavernosity with old fossilized passages was also discovered there by earthmoving works during the digging of the Kastelec tunnel where a network of caves opened up, over 500 m long. More than 20 other caves were opened up in the tunnel. It is expected that the cavernosity will be likewise high on the route of the railway line. The diameters of the passages will measure up to 10 m. The cavernosity with (sub-)vertical vadose shafts with diameters of up to 5 m will also be relatively dense. Great cavernosity is also expected along the fault zone at km 13.570.

Greater cavernosity is expected along the flysch slices at km 14.000, 14.240 and 15.450 (15.750) and everywhere along the contacts between Eocene Alveolina-Nummulites limestone, transitional beds and flysch.

During a medium to low water level, greater quantities of hanging water bodies are expected between km 11.280 and 11.900 in the area of the Beka-Ocizla cave system. During a high water level flows of water are active in this area above the height of the route of the railway line, which flood a few tens of metres above the planned height of the tunnel T 2.

Up to 10 caves/km with diameters of up to 10 m are expected.

Section K This section passes through the impermeable Eocene marly and flysch beds. Due to the carbonate binder in the transitional beds, smaller cavernosity is expected. Seeing that the route runs in the immediate vicinity of the contact of limestone with the transitional beds, karstification of the upper layer of marly beds is possible there. Due to the gently sloping overthrusting fault, the irruption of hanging water is possible throughout the entire route of this section.

Up to 5 caves/km with diameters of up to one metre are expected.

5 The Biology of Subterranean Habitats, Ecological Parameters and the State-of-the-Art of Subterranean Invertebrates in Selected Habitats

In karst carbonate landscapes both terrestrial and aquatic underground habitats have developed. There we can find organisms which are morphologically and physiologically adapted to specific conditions. Many of them are endemic, as they live in very limited areas. They represent an important source of information about the past and present ecological circumstances of the environment in which they live.

Caves are an important part of both the terrestrial and aquatic ecosystems in karst terrains, but are not the only components (Fig. 22). Among those that are less deep than caves are the so-called shallow subterranean habitats (SSHs; Pipan and Culver 2012). SSHs are aphotic habitats, such as seeps or seepage springs, the epikarst, the hyporheic and the mesovoid shallow substratum (MSS). The epikarst (Figs. 22, 23) is a perched aquifer and a major ecotone between surface water and cave water. Water is transmitted vertically either through conduits or small fissures to the phreatic zone. Lateral transmission occurs through poorly integrated lateral openings. Its principal characteristic is its heterogeneity, with many semi-isolated solution pockets whose water chemistry is also quite variable. It forms a more or less permanently saturated zone, close to the surface. The epikarst habitat contains many air-filled cavities as well, and terrestrial species are routinely collected from the epikarst. Stygobionts have been found in all of the locations where epikarst fauna has been sampled, yet the Slovenian

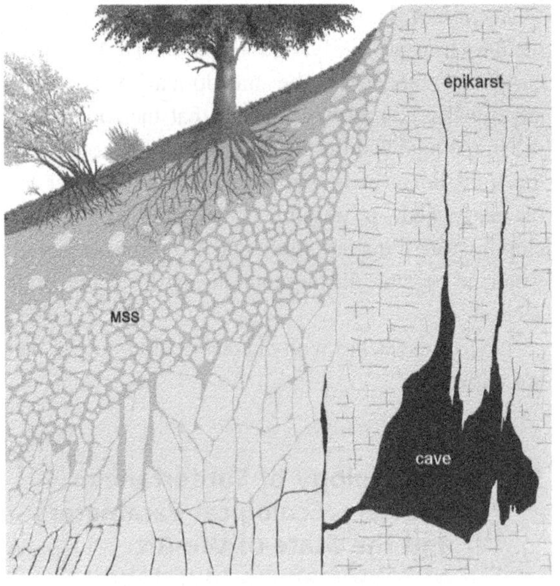

Fig. 22 The conceptual model of different types of subterranean habitats in the study area (from Latella and Sbordoni 2002)

assemblages are by far the most diverse (Pipan 2005). In all of the explored caves, copepod species were discovered to be the richest and most abundant fauna; many of them were epikarst specialists, but other taxa were also present.

The hypotelminorheic habitat or seepage spring (Figs. 24, 25), is:

- a persistent wet spot, a kind of perched aquifer fed by subsurface water in a slight depression in an area of a low to moderate slope;
- rich in organic matter;
- underlain by a clay layer typically 5–50 cm beneath the surface;
- with a drainage area typically less than 10,000 m^2;
- with a characteristic dark colour derived from decaying leaves, which are usually not skeletonized (Culver et al. 2006).

The habitat can occur in a wide variety of geologic settings anywhere outside of arid regions where there is a layer of impermeable sediment; however, it is probably less common in karst landscapes because the extensive occurrence of an impermeable clay layer would prevent the downward movement of water and the development of karst landscapes. The most common invertebrates found in seepage springs are amphipods (Fig. 26). Although some can be accidentals or stygophiles (species that can complete their life

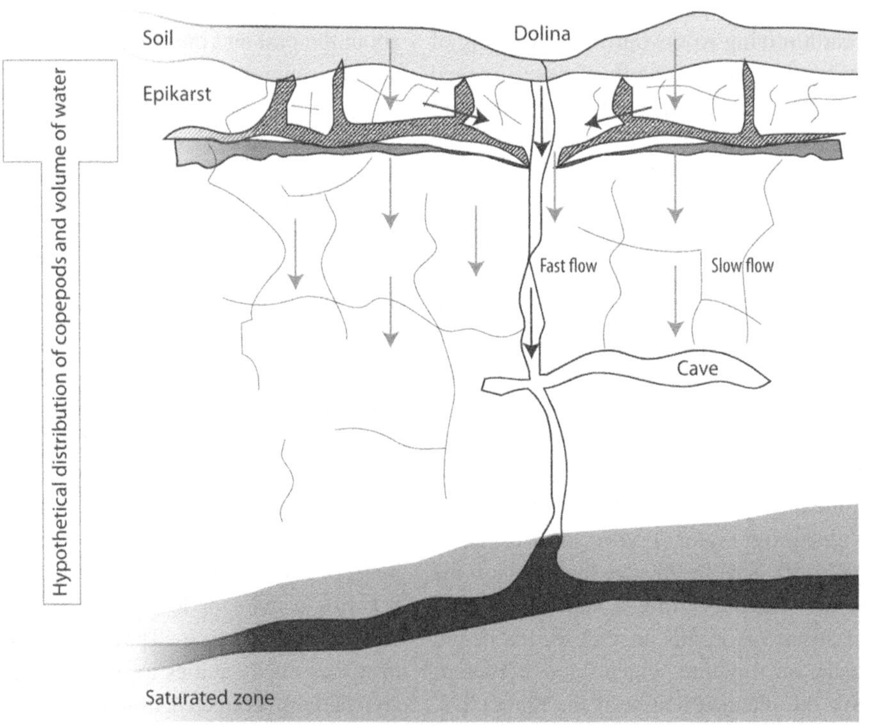

Fig. 23 The conceptual model of the epikarst. *Light arrows* indicate the direction of the slow water flow and *dark arrows* are the faster flow paths (from Pipan 2005)

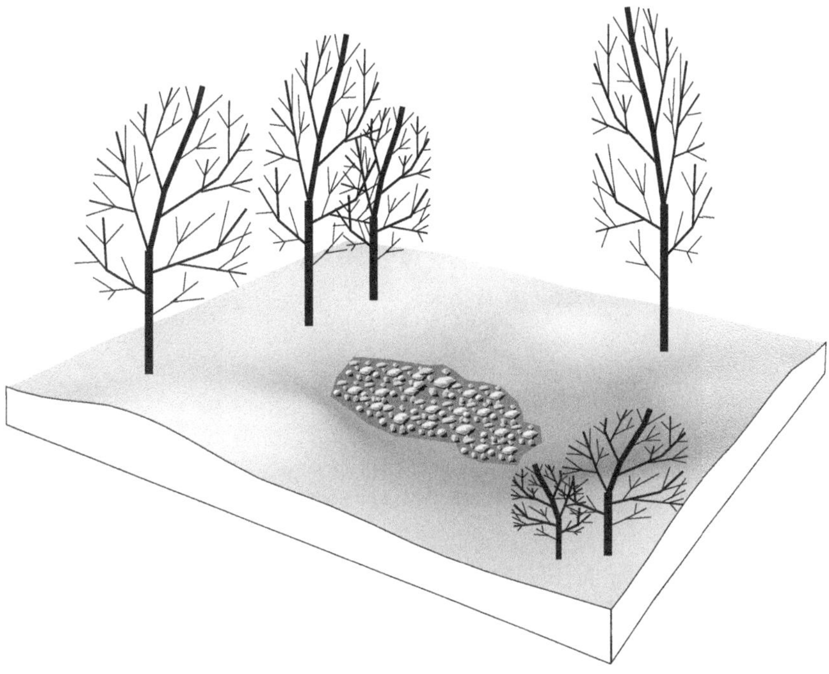

Fig. 24 A sketch of the hypotelminorheic (from Culver et al. 2006)

Fig. 25 Sampling of fauna in a seepage spring (hypotelminorheic)

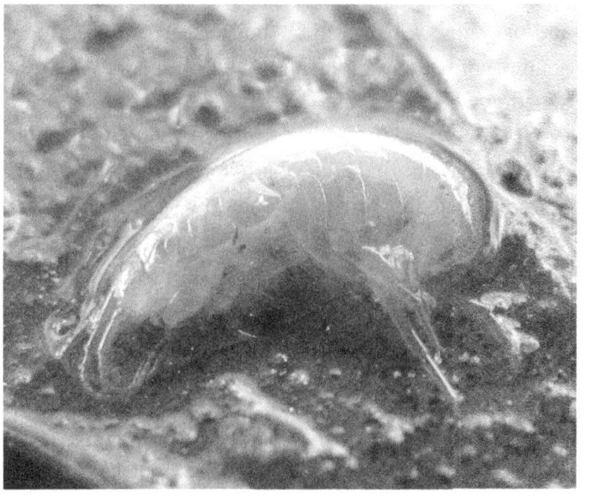

Fig. 26 A seepage spring is normally inhabited by the subterranean amphipod species of *Niphargus*

cycles in either subterranean or epigean habitats), many of them are stygobionts living exclusively in subterranean habitats. In addition, some can be found exclusively in seeps: they are hypotelminorheic specialists.

The range of terrestrial subterranean habitats is much more restricted than that of the aquatic subterranean habitats. But terrestrial subsurface habitats also occur outside the caves. Generally occurring at a depth of a few metres, this habitat was called *milieu souterrain superficiel* (MSS) in French, or *mesovoid*

Fig. 27 The photo of the exposed mesovoid shallow substratum (MSS) at a road cut where voids are filled with soil

Fig. 28 Sampling of fauna in the percolation water in the study area

Fig. 29 Sampling of fauna in the water pumped from the T 2-8 borehole

shallow substratum in English (Figs. 22, 27). The MSS is generally found in mountains in temperate zones but apparently not in the tropics where voids are usually filled with laterites and clay (Culver and Pipan 2013). This system of fissures and cracks filled with air and/or soil is an important habitat of many subterranean organisms. Beetles and collembolas are abundant and diverse in species.

Compared with related surface-dwelling species, subterranean animals are characterized by the absence or reduction of a series of features, particularly eyes and pigment (see Fig. 26). Subterranean animals are also characterized by elaboration of extraoptic sensory structures as well as other morphological, ecological, physiological, and behavioural adaptations. This represents a remarkable convergence among subterranean animals, the morphological aspects of which are called troglomorphy.

5.1 The Sampling Procedure and Results

In the study area, the fauna from the infiltration zone was sampled (Fig. 23). The habitat had to be sampled indirectly by collecting and filtering water from small pools which are filled with water from the trickles (Table 3). Stygobionts, i.e. species specialized for subterranean life, represent sink populations in the ecological terminology of source–sink populations. Nonetheless, they are important collection sites for the epikarst fauna, especially when the first survey of fauna is made in a particular cave and can yield a very diverse fauna. Biologically, epikarst is an interface layer, an ecotone between surface and subsurface water. As it is inaccessible to man, the direct sampling of epikarst habitats is not possible. Epikarst fauna must be explored indirectly by taking samples of percolation water (Fig. 28).

Boreholes in the karst underground offer the unique opportunity of direct access to the habitat developed in the unsaturated zone. In the boreholes, aquatic or terrestrial fauna that lives in the system of cracks and fissures below the surface, can be sampled directly. Preliminary studies have been carried out in the T 2-8 borehole (Fig. 29).

The most abundant organisms that are brought by the percolation water to the cave are copepod crustaceans (Fig. 30). Copepods are very diverse and are the

Table 3 Measured parameters in percolation water in the Osapska jama cave (5 February 2010)

Conductivity (μS/cm)	Temperature (°C)	pH	Volume (ml)
489	6.6	8.28	1,500

most numerous crustaceans in the aquatic community. Copepod habitats range from freshwater to hypersaline conditions, from the highest mountains and cold polar ice waters to the deepest ocean trenches. The sampling of any groundwater—cave stream, percolation water, deep phreatic water, hydrothermal water, interstitial or hyporheic water—yields many individuals of different copepod species. Such adaptation to different, also extreme environmental conditions is a result of their evolution; several invasions into groundwater habitats probably occurred in different time scales, which is what makes the group of Copepoda very diverse.

According to the ecological connectivity of species in the subterranean environment, subterranean animals are divided into:

- *Troglobionts or stygobionts* obligate, permanent residents of terrestrial or aquatic subterranean habitats. They are often troglomorphic with morphological and behavioural characters that are convergent in subterranean populations. Among them are subterranean snails, beetles, crustaceans and the well-known human fish *Proteus anguinus*.
- *Troglophiles or stygophiles* species that complete their entire life cycle in subterranean habitats or waters but can also complete their entire life cycle in the epigean domain or in surface waters. Some cave crickets, opiliones and insects, usually at the entrance part of caves, belong to this category.
- *Trogloxenes or stygoxenes* organisms with no affinities with subterranean habitats or groundwater systems but which occur accidentally in caves or alluvial sediments.

The karst underground domain, which at first glance seems to be averse to the existence and development of life, is inhabited by a large number of different organisms. The sampling of selected subterranean habitats provides insight into the rich subterranean biodiversity. Subterranean species can be used as bioindicators to help assess the overall ecological status of subterranean habitats. Data are supplemented with the measurement of physicochemical parameters (Table 3).

The Dinaric karst, and especially its Slovenian part, is a global hotspot for subterranean biodiversity. The subterranean fauna of Slovenia is among the richest in the world. Recent international studies (Culver et al. 2006) have found that a ridge (42–46°N in Europe and 34°N in North America) of high biodiversity occurs in temperate areas of high productivity and cave density. This may reflect a strong dependence of subterranean communities on long-term surface productivity because the subterranean fauna relies almost entirely on resources produced outside of caves.

For the Slovenian subterranean fauna, its research and conservation, the following statements are important:

- In Slovenia the research of subterranean life began nearly two centuries ago. The Slovenian populations

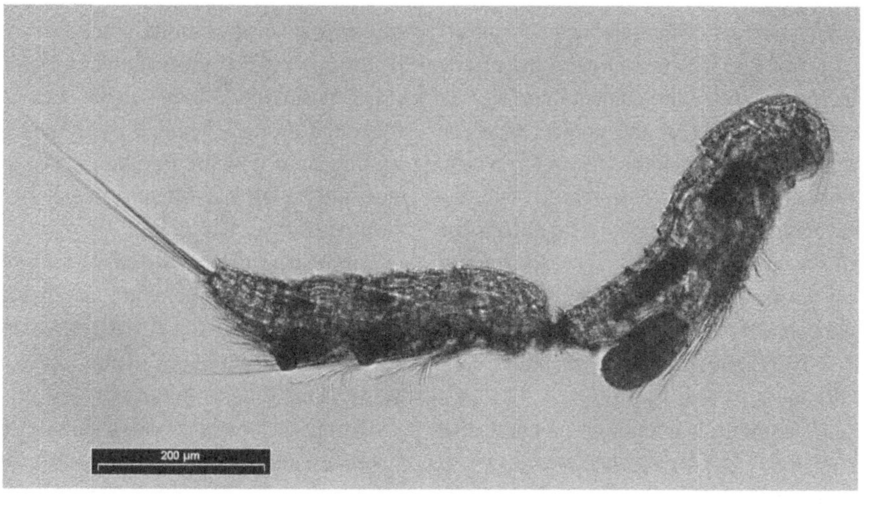

Fig. 30 Precopulatory mating individuals of epikarst copepods

of some subterranean species have become internationally important research models, especially for discovering their evolution and adaptation to new, extreme habitats, as well as their specialization and the genesis of biodiversity. Among the most important aquatic subterranean species are isopods (*Asellus aquaticus*) and blind cave amphipods (genus *Niphargus*).

- Subterranean environments in Slovenia are economically important as a source of drinking water, as a basis for tourism and as an important integral element of karst ecosystems in a broad sense. It is therefore important to explore them well and to learn about them. Only knowledge can form a good base for proper protection and conservation.
- Many subterranean species are specialists, sensitive to the transfer of organic and other substances in the groundwater. Due to the high level of ecological specialization, many subterranean species are used as bioindicators. The comprehensive impact of human activities on subterranean habitats can be assessed by studying subterranean fauna.

6 Cavernosity Assessment

Density of the registered caves, results of karstological supervision of the motorway construction in the karst, measurements in the profiles of the Črni Kal quarry, data from various geomechanical surveys, and knowledge of the development of the karst can be used to assess cavernosity.

There are 177 caves registered in the area in question. Along the route, three areas can be pointed out that have the highest density of cave entrances and cavernosity. The first is the area between Divača and Lokev where the main water route of the contact karst aquifer runs. The second is the area of the contact between flysch and limestone, along which the Beka-Ocizla cave system has developed. The T 2-9 borehole with the highly karstified core is located nearby. The third is the area of the shaft Brezno na Škrklovici where the longest cave opened up during the motorway construction in Slovenia (a cave over 500 m long in the Kastelec road-tunnel). The T 2-10 borehole, in which a larger cave opened up at 195 m, is located nearby.

For the risk assessment, it is enough to know that there were more than 350 caves discovered across 75 km of the motorway route in the karst. Two thirds of the caves are old including denuded, i.e. roofless caves. The rest are shafts.

It has been determined that the old caves and shafts are located in both Cretaceous and Paleocene limestones and in Lower Cretaceous bituminous dolomites. The majority of the shafts were discovered in Cretaceous limestone; relatively few of them were found in individual segments of Paleocene limestone. The latter is also perforated with numerous small pipes, which were observed in the cut beneath the Čebulovica hill. Only a selected few developed into larger caves. The railway line will also pass through flysch rocks. The flysch beds are mostly made up of water-impermeable marls and sandstones; the more permeable beds can be found within the flysch sedimentation. These are nummulitid breccias and limestones. The properties of the rock are less important for the creation and formation of caves than the rock stratification and its fracturedness; i.e., above all the characteristics of certain bedding planes (Knez 1996) and fissures and their networks of forming cavities alongside them under various conditions of the development of the aquifer.

A high density of caves is also present near the Karst Edge; measurements in the excavations of the Črnotiče quarry have shown 3.9 % cavernosity of the upper 19 m of Eocene limestone.

These are the oldest remnants of speleogenesis in this region. Similar passages are expected throughout the entire unsaturated zone, which were created during various speleogenetic periods. These passages can measure over 10 m in diameter.

Based on the diameters of the cores on the 11, so to speak, spot surveys in the boreholes over a distance of 10 km, then from determined karstification, from the caves, fissures and zones in the karst that are open to water and through which drilling mud flowed out up to six times in certain boreholes, we can deduce the incredibly high karstification level of the entire karst rock block.

Combining the results of various research leads to the conclusion that the entire area is highly perforated. It is believed that from 5 to 10 caves are likely to open up across one kilometre of the route, and 15 caves in the separately described areas.

During the construction, karstological monitoring is necessary and could in many cases enable much faster overcoming of obstacles during the construction.

Karst Waters

1 Hydrogeological Characteristics of the Karst Section of the Planned Railway Route

An imbricate overthrust structure is characteristic of the broader area of the planned railway line between Črni Kal and Divača. Among the carbonate rocks which are mainly limestones of the Upper Cretaceous and Palaeocene age outcrop zones of Eocene flysch (Fig. 1).

Hydrogeologically speaking, carbonate rocks are defined as highly permeable karst aquifers in which the underground water flow is predominant. Due to the thin layer of soil and the fissured and karstified upper part of the karst rock, the infiltration of precipitation water into the karst aquifer, along with substances dissolved within it, is fast, while the possibility of filtering harmful substances is slight. Even more direct is the introduction of surface water into the system of highly permeable karst channels through the sinking of surface rivers from the non-karstic edge at the contact with the karst. The channels and extended fissures in the karst aquifer enable a very fast flow over great distances. The results of this are short water retention times in the underground and consequently poor self-cleaning ability and great vulnerability of karst springs, towards which the groundwater flows and from which it then flows out onto the surface. Some of the water is stored in the less permeable zones of the aquifer for a longer period of time. This results in the accumulation of potential contamination, which the water can squeeze out of the system even after a longer period of time under suitable hydrological conditions. Therefore, karst aquifers have a characteristic heterogeneous structure, within which larger karst channels and extended fissures with the role of primary drainage paths intertwine with the less permeable zones through which the water flows slowly.

Flysch consists of very poorly permeable marls and sandstones, with beds of more permeable breccia and limestones in between. The flysch layers between karst aquifers present local hydrogeological barriers, along which water may accumulate; at the contact, increased inflows may appear. The flysch rocks on the surface enable surface flows that sink into the underground at the contact with the karst. On the other hand, flysch also represents a hydrogeological barrier, along which underground karst waters flow out onto the surface in karst springs.

1.1 Karst Water Sources

The main paths of the groundwater flow from the area of the planned route are directed towards the springs of Timava, Rižana, Osapska reka and Boljunec (Fig. 1). Numerous other springs which emerge at the contact of the karst aquifer with the impermeable flysch or represent a direct outflow into the sea are smaller and of a more local character. Here follows a description of the basic characteristics of the most important water sources.

1.1.1 Timava Spring and Other Springs in the Gulf of Trieste

The groundwater from the Classical Karst flows out onto the surface in Italy in numerous springs in the Gulf of Trieste. The largest of these is the Timava spring (Fig. 2), which has three main outlets.

M. Knez et al. (eds.), *The Beka-Ocizla Cave System*, Cave and Karst Systems of the World,
DOI: 10.1007/978-3-319-04456-9_4, © Springer International Publishing Switzerland 2015

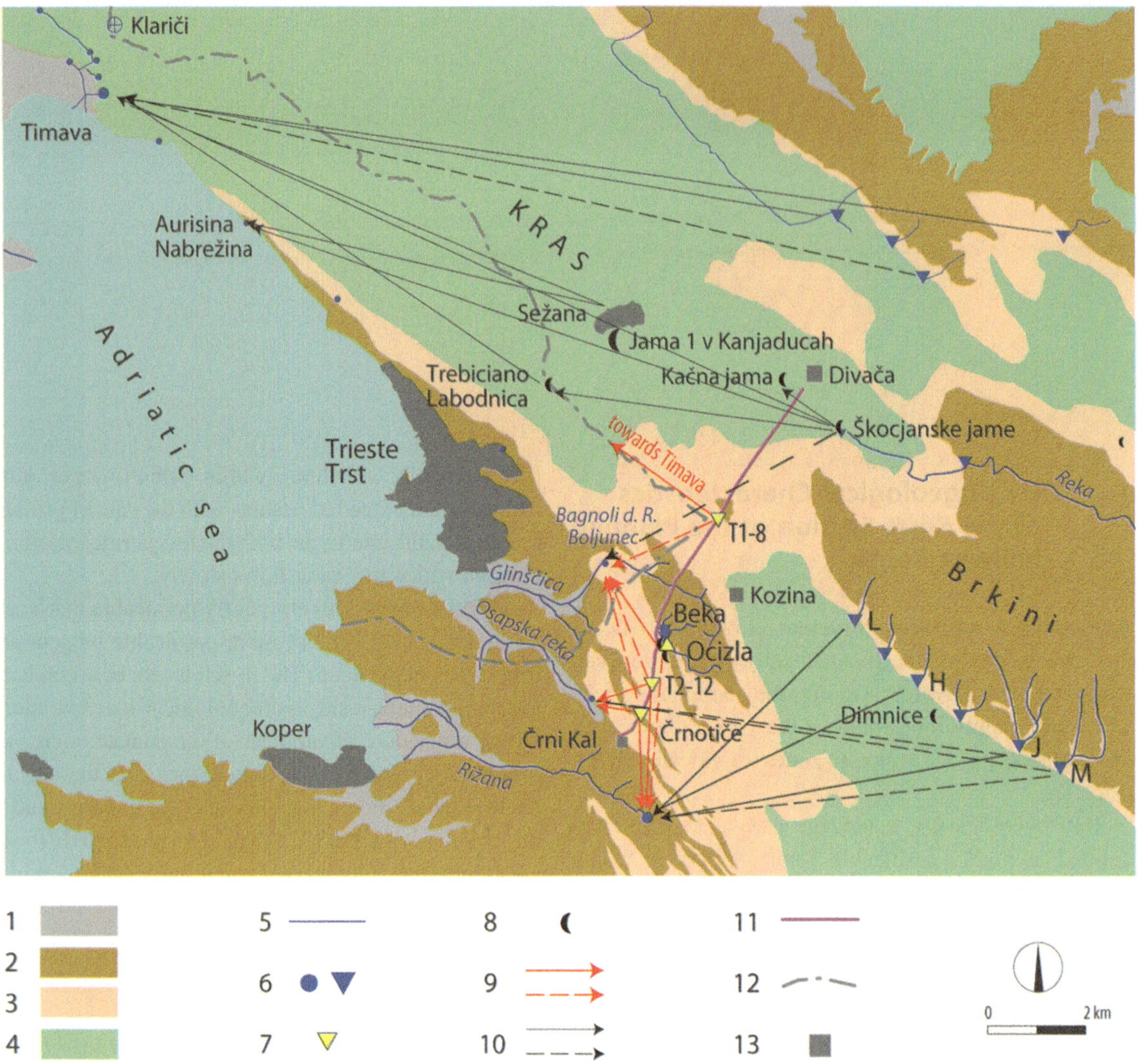

Fig. 1 Hydrogeological map of the broader area of the planned railway route. **1** Highly permeable rocks with intergranular porosity, **2** Very low permeable Eocene flysch, **3** Highly permeable Tertiary limestone and dolomite, **4** Highly permeable Cretaceous limestone and dolomite, **5** A surface stream, **6** A spring, ponor, **7** An injection point in tracer tests on the route of the railway, **8** A cave, **9** Main and secondary direction of groundwater flow shown by tracer tests on the route of the railway, **10** Main and secondary direction of groundwater flow shown by previous tracer tests, **11** The planned railway route, **12** The state border, **13** A settlement

According to data for the 1972–1983 period (Cività et al. 1995), the total discharge ranges between 9.1 and 127 m^3/s; the mean discharge is 30.2 m^3/s. In the past, this spring was captured for the water supply of Trieste; however, due to the decrease in quality, it was replaced by pumping water from the granular aquifer of alluvial sediments by the Soča/Isonzo River.

Numerous smaller springs are located deeper inland at the altitudes between 0.4 and 12 m; of special interest are the submarine springs along the coast towards Trieste. The most important among them is Brojnica near Nabrežina/Aurisina, which was captured for water supply from 1857 to 1977. The total mean discharge of all the smaller springs has been estimated at around 6 m^3/s.

1.1.2 The Klariči Pumping Station

In the area of the Classical Karst the most important water source is the pumping station in Klariči (Fig. 3), from which five karst municipalities with

Fig. 2 The Timava spring

approximately 22,500 inhabitants are being supplied with drinking water since 1984. During low water levels, some of the water is given to the Rižana water supply system, which does not have sufficient resources to cover its needs during summer droughts. At the bottom of the doline, at an altitude of 16 m, three wells have been drilled at a depth of 70 m, from which up to 250 l of water can be pumped per second. The water source is protected with a decree on the protection of waters; the protection zones currently in force do not reach the area of the railway route. Nevertheless, the Klariči pumping station was especially pointed out as the most important water source for the area of the Classical Karst.

1.1.3 The Rižana Spring

The Rižana spring is the most important karst spring in the Slovenian Littoral and has been captured for the water supply of this area since 1935 (Fig. 4a, b). Based on the conducted basic hydrogeological research and numerous tracer experiments (Krivic et al. 1987, 1989), the recharge area of the spring was estimated at 247 km^2. The main part is karstic, but the spring is also supplied with water from the flysch Brkini hills, which sinks at the contact with carbonate rocks. The Rižana spring emerges at the contact of the carbonate aquifer with the very poorly permeable flysch rocks across which the Rižana River flows into the Adriatic Sea. The discharges of the Rižana spring range from 30 l/s to 91 m^3/s; the mean discharge is 4.3 m^3/s.

The Rižana water source is protected with the Decree on Determining the Drinking Water Protection Area for the Aquifers of Rižana, which defines three protection zones. In the southern part, the planned route intersects the broader water protection area (zone III), within which the construction of the railway line

Fig. 3 The Klariči pumping station

Fig. 4 The Rižana spring at low (**a**) and high water levels (**b**)

is permitted if it is in accordance with the national or municipal site plan, adopted pursuant to the spatial planning regulations, if a comprehensive assessment of the impact on the environment has been performed for this site plan in accordance with the regulations governing the comprehensive assessment of the impact on the environment, and if protective measures are implemented due to the impact of construction on the water regime and status of the body of water, which show, based on the results of the analysis of the contamination risk, that the risk of contamination from construction is acceptable.

1.1.4 The Osapska Reka Spring

Above the Osp village, an intermittent spring of Osapska reka flows from a karst cave (Fig. 5). The entrance to the 1,200 m long and 49 m deep cave is located 105 m above sea level. In the cave, at the end of the passage Spodnji rov or Glavni rov, there are two permanent water pools, which have most likely lost contact with the water table when the water level was lowered (Novak 1964/65). The spring is only active after heavier rain; during a high water level, the discharges can amount to several m^3/s. Even before the water flows out of the cave, numerous springs become activated in the riverbed below the entrance.

Fig. 5 The Osapska reka spring at high water levels

1.1.5 The Boljunec Springs

On the Italian side, there are several springs in the Valley of the Glinščica/Val Rosandra at the altitude around 70 m in the village of Boljunec/Bagnoli della Rosandra. They are known by the common name of Boljunec. The springs Na placu and Pri pralnici (Fig. 6a, b) are permanent; occasionally, as an overflow of the Pri pralnici spring, water also flows out of a karst cave. It is called the Jama spring. During a low water level, the total discharge of the springs amounts to merely a few litres per second; during high waters it can exceed 2 m^3/s. The water from the Pri pralnici spring has been routed to a nearby fish farm.

1.2 Sinking Streams and Water Caves

Also connected with the alternation of differently permeable rocks is the phenomenon of ponors in which the surface waters from flysch sink at the contact with karstified carbonate rocks and then flow through the underground towards the karst springs. The closest to the route are the ponors of the Beka-Ocizla cave system, which together drain a flysch surface area of 3.5 km^2, across which several smaller, intermittent streams flow (see Fig. 1).

The surface Reka River sinks into the cave Škocjanske jame approximately 3.5 km east of the railway route. Its minimum discharge is 0.18 m^3/s, while its mean discharge is 8.26 m^3/s (Kolbezen and Pristov 1998). During very high water levels, its discharge can increase to over 300 m^3/s. As has been proved with numerous tracer experiments which are presented in

Fig. 6 The permanent springs Na placu (**a**) and Pri pralnici (**b**) in the village of Boljunec

greater detail in the Sect. 2, these waters then flow away underground towards the Timava spring and to a smaller degree towards other springs in the Gulf of Trieste. The possibility of a connection with the Boljunec spring has not been ruled out.

Numerous sinking streams are found at the foothills of the flysch Brkini hills. Most of them flow away underground towards the Rižana spring, whereas in some cases tracer experiments have established a possibility of the flow towards the Osapska reka spring.

1.3 Water Table

In order to provide a general presentation of the characteristics of the water table fluctuation in the karst aquifers in question, we present some basic data on the levels of water in karst caves and in karst springs. In the Škocjanske jame cave, the lowest water level is around 210 m above sea level, whereas in the Kačna jama cave near Divača it is 156 m above sea level (see Fig. 1). Towards the west, the water table drops towards the Labodnica/Trebiciano cave in Italy to the altitude of 12 m. During high waters the level in Škocjanske jame reaches 345 m, in Kačna jama 260 m and in Labodnica 112 m. The underground flow in this segment of the karst aquifer is directed towards the Timava spring which is situated merely some 0.5 m above sea level.

In the sink caves at the foothills of the Brkini hills, the permanent water flow can be observed only in the Dimnice cave at the altitude of 474 m. The Rižana spring is situated at around 70 m above sea level; during a low water level, the water in the cave Osapska jama drops below 50 m, whereas during high water levels it springs out of the cave at the altitude of 105 m. Based on the data, it can be concluded that the level of karst water in this area slopes steeply towards the southwest and, due to the imbricate structure of alternating flysch water barriers between karstified limestones, also in grades (Habič 1985). In Boljunec, the water springs at around 50 m above sea level.

2 Directions and Characteristics of the Underground Water Flow

2.1 The Review of the Results of Previous Tracer Tests

An important additional information on the courses and characteristics of the underground water flow in the karst can be obtained by applying a tracer test method, which monitors the transfer of the injected tracer (a water-soluble substance, harmless to the environment, which can be detected in very low concentrations) through the aquifer. The ascertained directions and flow velocities of the water and the recovery of the tracer in individual springs, in which the tracer appeared, provide a great deal of information on the way the water flows through the karst under given hydrogeological conditions and indirectly on the transfer of potential contamination.

The first tracings were conducted in the Classical Karst already in the early twentieth century and were followed by many others, which showed that the Reka River flowed from the Škocjanske jame to the Timava spring. The tracer was also detected in various water caves between these two extreme points of the underground flow of the Reka River (Fig. 1 and

Table 1 Results of the tracer tests

Injection point	Date	Tracer	Proved connection	Apparent flow velocity (m/h)	Reference
Reka River (Škocjanske jame)	23/12/1907	LiCl	Timava, Brojnica	162 128	Timeus (1928)
Reka River (Škocjanske jame)	28/1/1913	Uranine	Labodnica	97	Timeus (1928)
Reka River (Škocjanske jame)	3/7/1962	Uranine, tritium	Labodnica, Brojnica, Timava, Boljunec	200 50 86 104	Mosetti (1989)
Reka River (Škocjanske jame)	4/9/2006	Uranine	Jama 1 v Kanjaducah, Timava	164 51	Peric et al. (unpublished)
Sežana landfill	20/4/2005	Uranine	Timava, Brojnica	39 19	Kogovšek and Petrič (2007)
Ločica (L on Fig. 1)	10/4/1985	Uranine	Rižana	101	Krivic et al. (1987)
Jezerina (J on Fig. 1)	13/5/1986	Rhodamine	Rižana, Osapska reka	30 30	Krivic et al. (1989)
Male Loče (M on Fig. 1)	13/5/1986	Uranine	Rižana, Osapska reka	30 30	Krivic et al. (1989)
Hotiški potok (H on Fig. 1)	13/5/1986	Phages	Rižana	26	Krivic et al. (1989)
Beka-Ocizla cave system	23/5/1908	Fuchsine	Boljunec	no data	Timeus (1928)
Beka-Ocizla cave system	29/3/2001	Uranine	Boljunec (Pri pralnici, Jama), Rižana	33 29	Kogovšek and Petrič (2004)
Črnotiče	1/12/2009	Uranine	Rižana, Osapska reka, Boljunec (Pri pralnici)	22 33 10	Petric and Kogovšek (2011)
T 2–12 borehole	18/11/2010	Uranine	Rižana, Osapska reka, Boljunec (Pri pralnici)	62 23 48	
T 1–8 borehole	18/11/2010	Amidorhodamine G	Boljunec (Pri pralnici)	61	

Table 1). The determined apparent flow velocities (with regard to duration of the transfer of the tracer and the linear distance between the injection spot and the spot in which the tracer appeared), which depend on the hydrological conditions amount to around 40 m/h during the lowest water levels, from 80 to 90 m/h during low waters, from 109 to 164 m/h during medium-high waters, and exceed 300 m/h during high waters. Such high velocities are characteristic of the flow through karst channels.

In April 2005, we conducted the tracing from the surface in the area of the Sežana waste disposal site (Kogovšek and Petrič 2007). The connection with the Timava spring and with Brojnica spring near Nabrežina/Aurisina was confirmed. The apparent flow velocity in the entire section where the tracer first had to pass from the surface through some 300 m thick vadose zone and then flowed through the underground channels towards the Timava spring was 39 m/h. A part of the tracer, which was stored in the vadose zone, continued to be rinsed out for a longer period of time, after it rained. The vadose zone thus slowed down the water flow containing the tracer, but it did not represent a greater barrier. It has been demonstrated that when injecting a tracer on the surface or by analogy when the contamination appears on the surface, the course of the transfer of the substance through the karst aquifer is decisively affected by the prior saturation of the vadose zone and the distribution and intensity of precipitation.

Between 1985 and 1987 (Krivic et al. 1987, 1989), numerous tracer tests were conducted in the sinking streams at the foothills of the Brkini hills with the intention of determining the recharge area of the Rižana spring; certain connections with the Osapska reka spring were also established.

Timeus (1928) reports that in March 1908 the connection of the Beka-Ocizla cave system with the springs in Boljunec was proved with tracing.

The tracer tests described above were not enough to more precisely determine directions and characteristics of the flow of water away from the area of the planned railway route. Therefore, in 2001, 2009 and 2010, three more tracer tests were conducted which are described in greater detail in the next Sects. 2.2–2.5.

2.2 Description of the Used Tracing Method

The Slovenian Environment Agency (ARSO) carries out regular measurements of discharges of the Rižana River at the Kubed II station. In 2009 and 2010, we set up the Eijkelkamp water level data logger at the Osapska reka spring, which enabled storing of data about the levels in 30-min intervals; the discharges were then assessed, based on a comparison of older data on the discharges of the Osapska reka and Rižana springs.

The Eijkelkamp water level data logger was also placed at the common discharge of the Boljunec spring during all three tests. In parallel, we measured the total discharge with the Ott C20 current meter during different water levels (Fig. 7a, b). Simultaneously, we used the current meter to measure the discharge of the Na placu spring, which differs from the springs Pri pralnici and Jama in its physical and chemical characteristics, and in which the appearance of the tracer was not anticipated in light of the results of the previous research. These values were deducted from the total discharge and thus the total discharge of the Pri pralnici and Jama springs, in which the injected tracers appeared was obtained. By comparing the measured values, we obtained the rating curve of the dependence between the water levels and discharges, and based on the obtained equations for all the measured levels, we calculated the discharges.

In 2009 and 2010, we wished to monitor precipitation conditions in the area in question, so we placed the Eijkelkamp e+ rain logger near the Rižana spring, which logged the precipitation quantity in 15-min intervals.

The tracing was conducted with fluorescent tracers, uranine and amidorhodamine G. Different methods of injecting the tracer were applied: into the water flow of a sinking stream, into a well-permeable fissure on the surface and into two boreholes.

Samples were regularly taken at the Rižana, Osapska reka and Boljunec springs, while control samples were taken at certain smaller springs in the area. Blind samples were taken beforehand; afterwards, the samples were taken mostly with the ISCO 6700 automatic samplers and, in some places, manually. The frequency was constantly adapted to the hydrological conditions and the current results of the tracer test; the most frequent sampling took place every 4 h. During both recent tracings in the Rižana spring, we conducted parallel measurements of the fluorescence directly at the spring with the LLF-M field fluorometer in 30-min intervals (Fig. 8a, b).

In order to establish the presence of the tracers, we analysed the water samples with the PERKIN ELMER LS 30 luminescence spectrometer; the uranine at $E_{ex} = 491$ nm, $E_{em} = 512$ nm, and the amidorhodamine G at $E_{ex} = 531$ nm, $E_{em} = 552$ nm. The analyses were performed shortly after sampling and later on several times after decantation. The samples showed a high level of turbidity, especially in the samples taken from the Osapska reka spring where particles were deposited on the pumping pipe of the automatic sampler in the form of a light yellow-brown compact coating. The blind samples from the Rižana and Osapska reka springs taken prior to injection indicated only slight fluctuations of the signal when determining uranine,

Fig. 7 Placement of the water level data logger at the common flow of the Boljunec springs (**a**) and measurement of the total discharge with the current meter (**b**)

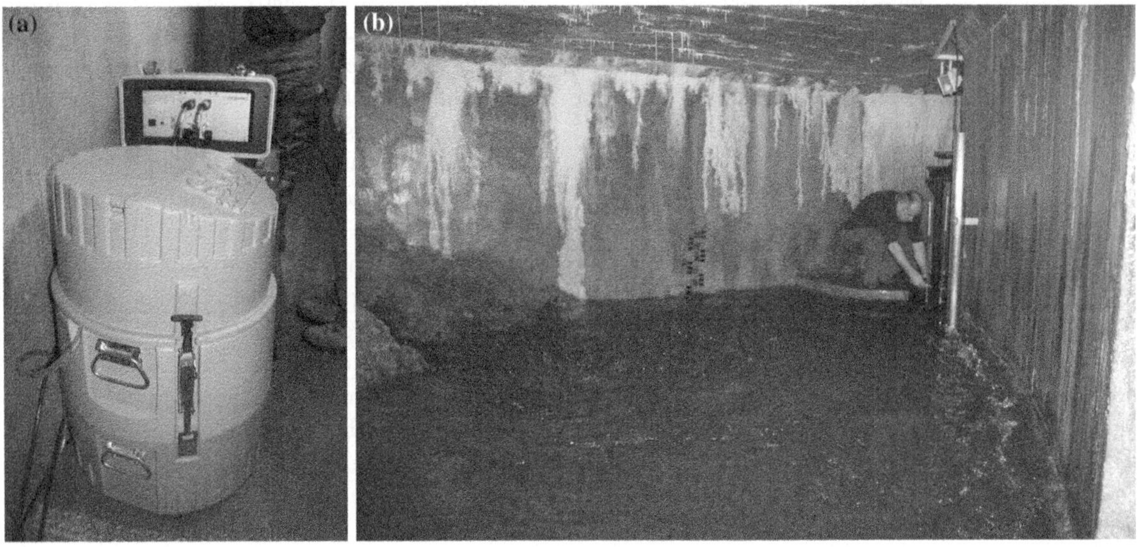

Fig. 8 The ISCO 6700 automatic water sampler and the LLF-M field fluorometer (**a**) and setting up of probes and pumping pipes in the reservoir at the Rižana spring in 2010 (**b**)

whereas in the Pri pralnici spring in Boljunec greater fluctuations were recorded. Greater fluctuations were also recorded in the Rižana spring when determining amidorhodamine G. All of these greater fluctuations most likely reflect the contamination from the population; in the case of Boljunec, they can also be explained with the use of uranine in this area for other purposes.

2.3 Tracing with the Tracer Injection into a Stream that Sinks into the Beka-Ocizla Cave System

On 29 March 2001, we injected a solution of 3 kg of uranine into a stream that sinks into the Beka-Ocizla cave system (Kogovšek and Petrič 2004).

Fig. 9 Injection of uranine into a stream that sinks into the Beka-Ocizla cave system

The tracing was conducted during a medium-high water level which gradually dropped towards the end of the experiment (Figs. 9 and 10). The tracer appeared distinctly in the springs in Boljunec, namely in the permanent Pri pralnici spring and in the Jama overflow spring, but not in the nearby Na placu spring. Parallel chemical analyses also showed that the Na placu spring differs from the Pri pralnici spring by the electrical conductivity and the content of carbonates, calcium, magnesium, chlorides, nitrates and o-phosphates. The Na placu spring contained a 4 times higher content of sulphates. It can be concluded that despite the distance of only a few tens of metres, the two springs has different recharge areas.

The uranine presence was not detected in the Osapska reka spring, which was active until the end of April. The tracer appeared in the Rižana spring with a greater delay and less distinctly. The first appearance of uranine in Boljunec was detected 84 h after injection; thus, according to the first appearance of the tracer, the apparent maximum flow velocity amounts to 42 m/h. The maximum concentration of uranine was measured 108 h after injection; the apparent dominant velocity amounts to 32.7 m/h. In the Rižana spring, the first brief appearance of the trace of uranine was measured about 10 days later; a more noticeable increase in concentration was determined as late as 50 days after injection. The apparent dominant velocity of the tracer in the direction of the Rižana spring is thus merely 6 m/h. Based on the measured uranine concentrations and the estimated discharge of the springs, it has been calculated that until early June 2001, in a bit over 2 months, around 91 % of the injected tracer flowed out through Boljunec, and around 2 % of the tracer through the Rižana spring.

2.4 Tracing with the Tracer Injection into a Fissure on the Surface Near Črnotiče

During a high water level on 1 December 2009, we injected a solution of 3 kg of uranine into a well-permeable fissure near the village of Črnotiče and flooded it with 2.5 m^3 of water from a fire cistern (Fig. 11a, b).

The tracing has shown that waters from the injection area flow mainly in the direction of the Rižana spring (87 % of the tracer flowed out) and Osapska reka spring (11 %); however, the flow towards Osapska reka is faster during higher water levels (Fig. 12). A smaller portion of water flows out a bit more slowly towards the Pri pralnici spring in Boljunec. Under the conditions of a relatively well-saturated vadose zone, prior to injection and high gradients from the injection spot to individual springs, the maximum velocities (with regard to the first appearance of the tracer) between 7.9 and 57.4 m/h and the dominant velocities v_{dom} (with regard to the time until the maximum concentration was reached) between 3 and 32.7 m/h were determined. The tracer flowed the fastest in the direction of the Osapska reka spring with a gradient almost two times higher than in the direction of other springs; it flowed the slowest towards the Boljunec spring. The flow towards the Rižana spring was relatively fast (v_{dom} = 21.7 m/h) and two longer lasting tracer breakthrough curves were created. The preliminary findings on the flow of water from the Beka-Ocizla cave system towards the spring of Boljunec can

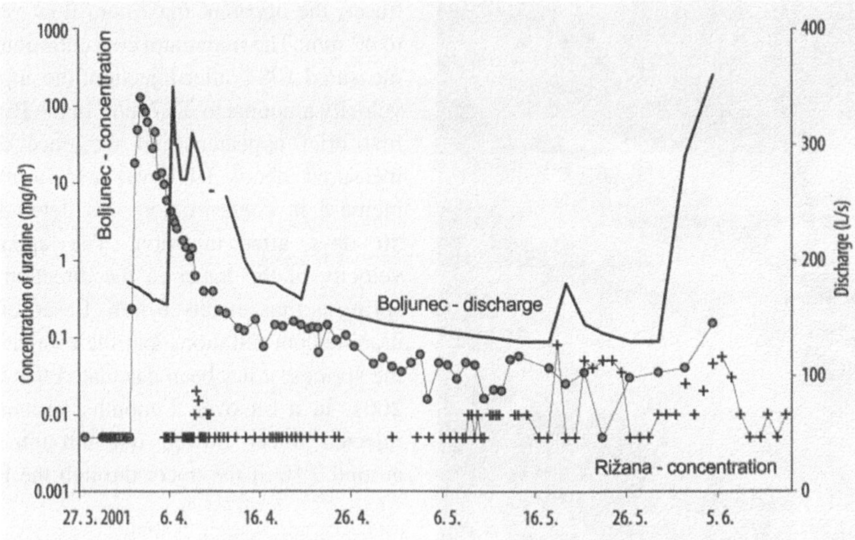

Fig. 10 Tracer breakthrough curves in the Pri pralnici (Boljunec) and Rižana springs and the discharge of the Boljunec springs

Fig. 11 Injection of uranine into a fissure near the Črnotiče quarry (**a**) and additional spray irrigation of the injection spot (**b**)

now be supplemented with the finding that slightly south from there the main course of the groundwater flow is towards the south and west, i.e. towards the Osapska reka and Rižana springs.

2.5 Tracing with the Tracer Injection into the T 1–8 and T 2–12 Boreholes

After two tracer tests were conducted, the issue of two karst watersheds in the area of the planned railway route remained open. In the southern part of the karst area, between the flysch zone near Glinščica and Divača, the possibility of an outflow towards the Boljunec spring remained undetermined; the same can be said for the watershed between Boljunec and Rižana in the bottom section of the route above Črni Kal (see Fig. 1). With intention of obtaining the required additional data, we conducted a combined tracer test using two tracers under favourable hydrological conditions of a high water level at the end of 2010. On 18 November 2010, we injected a solution of 305 g of amidorhodamine G (Fig. 13) into the T 1–8 borehole, and a solution of 4 kg of uranine into the T 2–12 borehole (Fig. 14a, b). In both cases, we flooded the boreholes and their surroundings with 6 m^3 of water from the fire cistern each, before and after injection.

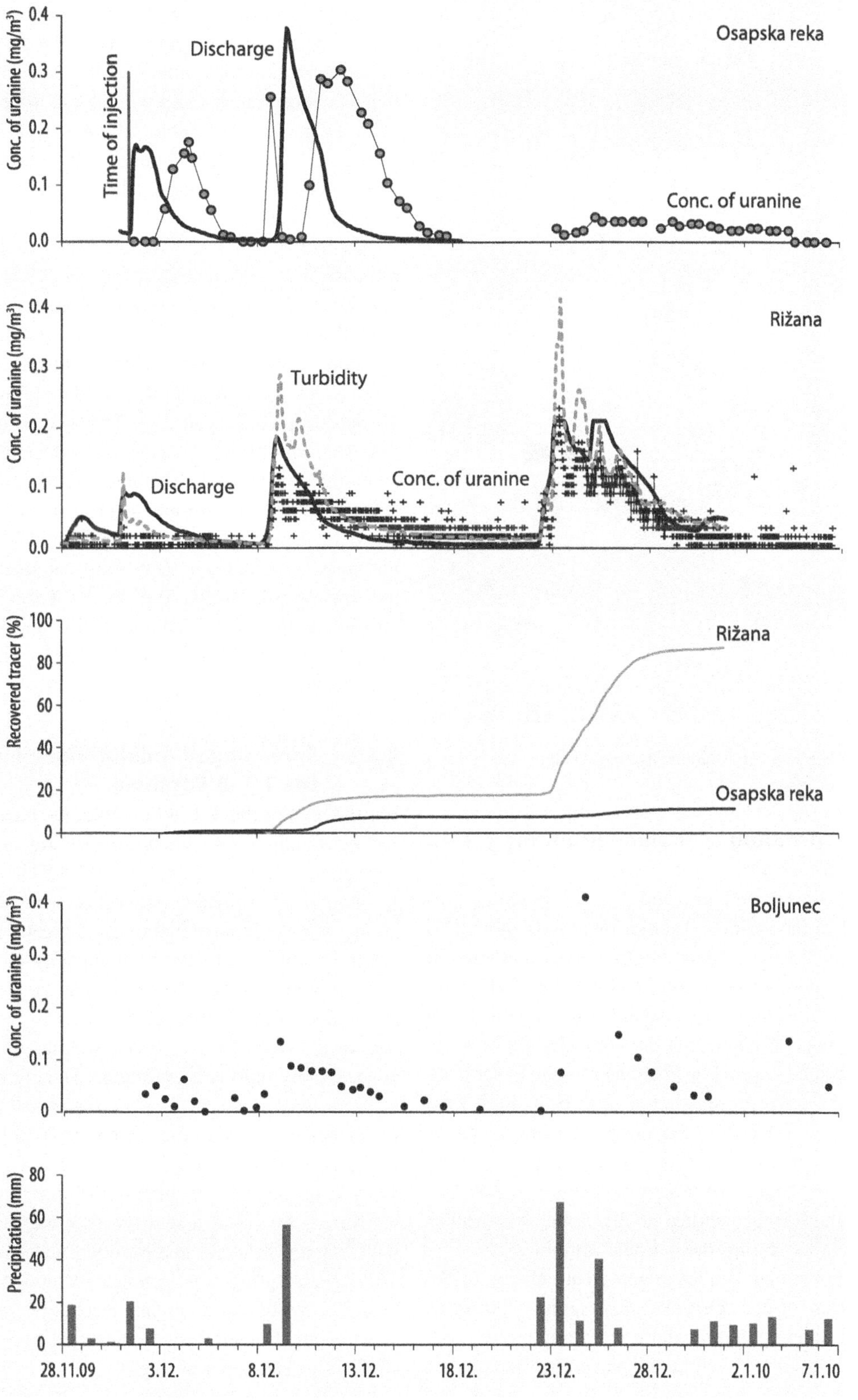

Fig. 12 Concentrations and shares of recovered tracer in the Osapska reka and Rižana springs and discharges of the two springs, concentration of uranine in the Pri pralnici spring (Boljunec), measured precipitation

Fig. 13 Injection of amidorhodamine G into the T 1–8 borehole

2.5.1 Spreading of Uranine (from the T 2–12 Borehole)

The uranine injected into the T 2–12 borehole was detected in the springs Osapska reka, Rižana and Pri pralnici in Boljunec when the discharges increased in the early morning hours of 22 November 2010 after heavy precipitation. It appeared most distinctly in Osapska reka; its concentrations were slightly lower in the Pri pralnici spring in Boljunec; even lower concentrations were present in the first peak in Rižana (Fig. 15 and Table 2). After the next stronger precipitation event, the Rižana spring kept reaching increasingly higher maximum concentrations of uranine. In the Na placu spring, no increased tracer values were recorded during the periodic sampling.

Due to the high discharge and great velocity of the water flow of the Osapska reka spring, problems occurred during pumping and a few samples are missing; in the sample taken on 22 November at 12 p.m., uranine was present in the concentration of 0.63 mg/m^3. Based on comparisons with other results, it has been concluded that uranine appeared in the spring earlier, in the early morning hours of the same day and at higher concentrations, around 1.1 mg/m^3 by our estimate (Fig. 15 and Table 2). When the discharge increased, the dilution caused the concentration to drop; when the discharge reduced, the concentration typically went up.

The calculated share of uranine, that flowed out through the Osapska reka spring by 7 January 2011, amounts to 33.2 % of the injected quantity; the portion that flowed out through the Pri pralnici spring, amounts to merely 1.6 %. Through the Rižana spring, 40 % of the injected quantity of uranine flowed out by 10 February 2011 (Table 2). This was followed by a drier period with only occasional lighter precipitation when the transfer of the retained uranine was minimal. It is presumed that a more intense transfer was triggered by abundant precipitation in mid-March when the springs were no longer being sampled. Through the springs of Rižana (up to 10 February 2011), Osapska reka (up to 7 January 2011) and Pri pralnici (up to 7 January 2011), a total of 76 % of the injected uranine flowed out.

2.5.2 Spreading of Amidorhodamine G (from the T 1–8 Borehole)

For the area of the T 1–8 borehole, we have predicted the possibility of an underground flow towards the springs near Boljunec and towards the Timava spring. In order to confirm this connection with the Timava spring, we would have had to use a greater quantity of the tracer, in light of the high capacity of the spring and the great distance; however, that would mean that the tracer might appear in the nearby springs near Boljunec, which have a much smaller discharge, in unacceptably high concentrations. Hence, we decided to use a smaller quantity of the tracer and to monitor its appearance only in the springs near Boljunec (due to the great dissolution, it could not be detected in the Timava spring). By calculating the share of the tracer recovered, we could assess whether we were dealing with a main or lateral underground water connection. The Rižana spring was sampled for amidorhodamine G too; however, due to greater discharges and distance and the small quantity of the injected tracer, the possibility of detection was slight.

Fig. 14 Injection of uranine into the T 2–12 borehole (**a**) and spray irrigation of the surface near the borehole (**b**)

The taking of samples of the springs in Boljunec showed that amidorhodamine G appeared in slightly higher concentrations in the Pri pralnici spring until 7 January 2011 (Fig. 16 and Table 2); no increased values were recorded in the Na placu spring.

In a month and a half, 7.3 % of the injected amidorhodamine G flowed out of the Pri pralnici spring. It has been concluded that most of the tracer flowed out in the direction of the Timava springs.

2.6 Characteristics of the Underground Water Flow

The presented results of the tracer tests (Table 1) provide the most reliable information about directions and characteristics of the groundwater flow in the karst. They also point to peculiarities of the functioning of karst aquifers in which waters flow out from a specific spot towards various springs in several

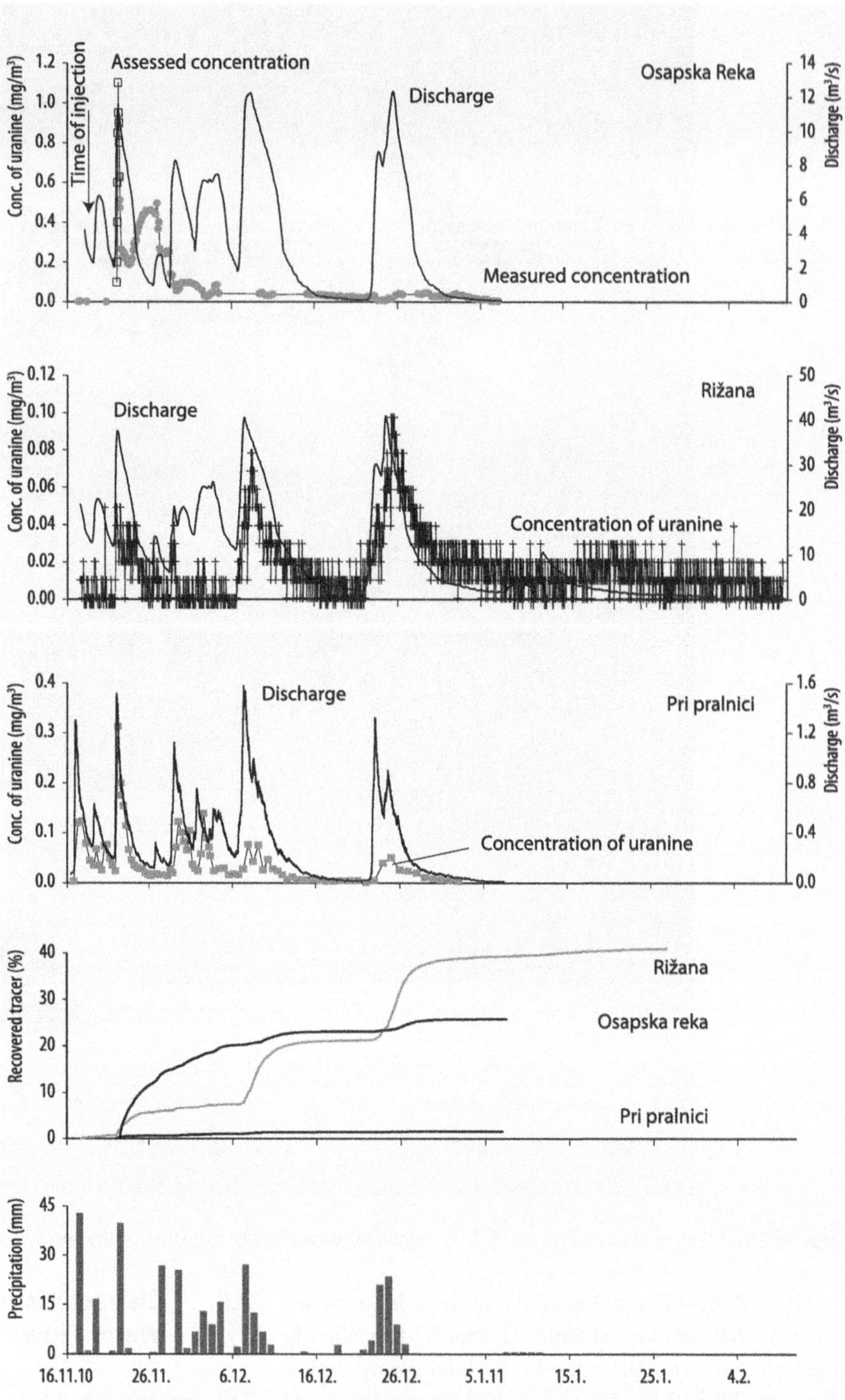

Fig. 15 Precipitation, discharges and tracer breakthrough and recovery curves for uranine injected into the T 2–12 borehole

directions; the underground pathways and the location of the watersheds between the catchment areas of individual springs change depending on the hydrological conditions.

The main directions of the groundwater flow can nevertheless be summed up for the area of the planned route of the railway line (Fig. 17). Thus from the area of the T 1 tunnel, north of the flysch

Table 2 Appearance of uranine and amidorhodamine G in the observed springs

Uranine		c_{max} (mg/m³)	t_{dom} (h)	v_{dom} (m/h)	M (kg)	R (%)
Rižana:	1st peak (22/11)	0.05	88	62	1.65	41.2
	2nd peak (8/12)	0.08	480	11.4		
	3rd peak (25/12)	0.10	883	6.2		
Osapska reka:	1st peak (22/11)	1.10*	93*	22.8*	1.33	33.2
	2nd peak (26/11)	0.46	188	11.3		
Boljunec-Pri pralnici	(22/11)	0.31	91	48	0.062	1.6
Amidorhodamine G						
Boljunec-Pri pralnici	(22/11)	0.11	89	61	0.022	7.3

c_{max} maksimum concentration (* assessed value), t_{dom} time of detection of the maximum concentration, v_{dom} dominant apparent flow velocity, *M* amount of recovered tracer, *R* share of the recovered tracer

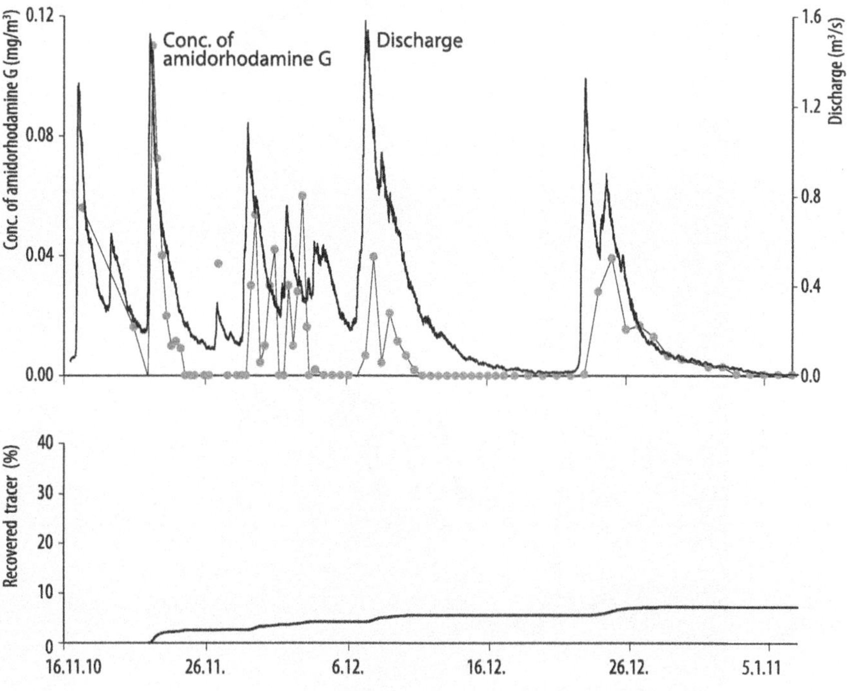

Fig. 16 Discharges and tracer breakthrough and recovery curves for amidorhodamine G injected into the T 1–8 borehole

zone of the Glinščica valley (from 0 to 6 km), which is composed of Upper Cretaceous and Palaeocene limestones, the underground flow is directed through the Classical Karst aquifer mainly towards the Timava spring; only a small portion of this water may appear in other springs in the Gulf of Trieste. Through the most permeable channels of the underground flow of the Reka River, the apparent velocity of the water flow ranges from 40 to 200 m/h during different water levels. These speeds are smaller through the less permeable zones within the aquifer. The tracer test in the area of the Sežana landfill has shown that the apparent velocity of the flow between the karst surface and the Timava spring during the spring's relatively low discharge was around 10 m/h.

Fig. 17 Directions of the underground water flow: for individual sections of the planned railway route (marked with *broken lines* in different colours) the most probable directions of the underground water flow are marked with *arrows* in different colours (*dark blue* for the Timava spring, *light blue* for the Boljunec spring, *yellow* for the Glinščica surface stream, *green* for the Rižana spring, and *orange* for the Osapska reka spring); a *solid line* represents the main direction (*thicker line* means larger share of flow in this direction) and a *dotted line* the secondary direction proved by the tracer tests; a *dashed line* represents possible connection which was not proved by the tracer test ▶

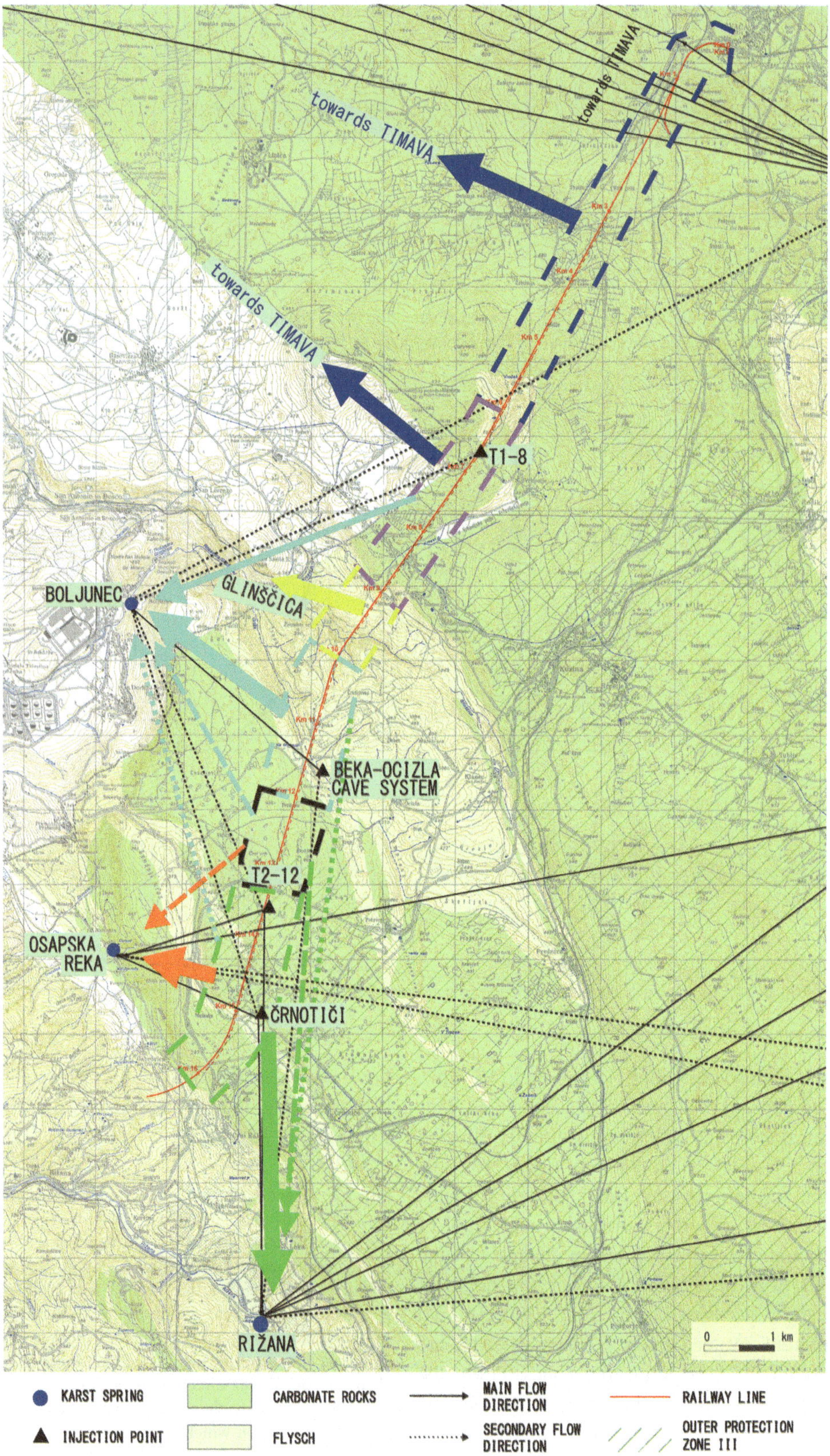
towards TIMAVA
towards TIMAVA
towards TIMAVA
T1-8
BOLJUNEC
GLINŠČICA
BEKA-OCIZLA CAVE SYSTEM
T2-12
OSAPSKA REKA
ČRNOTIČI
RIŽANA
0 1 km
KARST SPRING
INJECTION POINT
CARBONATE ROCKS
FLYSCH
MAIN FLOW DIRECTION
SECONDARY FLOW DIRECTION
RAILWAY LINE
OUTER PROTECTION ZONE III

A tracer test was conducted in the southern part of this section during a high water level by injecting the tracer into the T 1–8 borehole. Based on the results obtained, it has been hypothesized that the underground waters from this section (from km 6 to km 9) also flow mainly towards the Timava spring. The water flows to a smaller degree (about a tenth of the injected tracer was recovered) towards the Pri pralnici spring in Boljunec with velocities of around 60 m/h. In comparison with other methods of injecting the tracer, the tracing in the boreholes showed that the tracer was retained there the longest; hence, the flow velocities in this area could be somewhat higher.

From 9 km onwards, the T 1 tunnel runs across very poorly permeable flysch rocks and ends in a south portal above the valley of the Glinščica River, which flows on the surface into the Adriatic Sea.

The T 2 tunnel starts on the southern side of the Glinščica River; its initial segment is located in flysch rocks. The construction of the tunnel presents a possibility of the water flowing away to the contact with carbonate rocks. The Beka-Ocizla cave system is located on this contact; it has been proved that the waters from this system flow mainly towards the Pri pralnici spring in Boljunec. The tracer test conducted during a relatively low water level determined the apparent flow velocities between 30 and 40 m/h; over 90 % of the injected tracer was recovered. With a longer time lag and to a smaller degree (a few per cent), the tracer was also detected in the Rižana spring; therefore, the possibility of such a connection during higher water levels cannot be overlooked.

For the area between 12 and 13.5 km, the outflow towards the springs Pri pralnici in Boljunec, Osapska reka and Rižana can be assumed; yet, the priority directions cannot be determined more precisely. The portion of water that flows out towards individual springs also depends on the current hydrological conditions.

The results of two tracer tests refer to the area of the southern part of the T 2 tunnel between 13.5 km and the southern portal. Both have proved the main directions of the underground flow towards the Osapska reka and Rižana springs and confirmed the lateral connection with the Pri pralnici spring in Boljunec. During a high water level, the tracer first appeared in the Osapska reka spring (apparent velocity between 25 and 35 m/h) and somewhat later and in smaller concentrations in the Rižana spring (apparent velocity around 20 m/h). During the high water level, the portion of the tracer recovered was slightly higher in the Rižana spring than in the Osapska reka spring; based on the knowledge of the conditions there, it can be assumed that the portion of the outflow towards Rižana increases further when the water level drops. During a low water level when the Osapska reka spring dries up, virtually the entire underground flow is directed towards the Rižana spring. The outflow through the spring Pri pralnici in Boljunec is tied to a higher water level; the velocities are relatively lower and the portion of the tracer recovered amounts to merely a few percent. It is possible that these velocities are somewhat higher, since, based on a comparison of the results of the conducted tracings, it has been deduced that, due to the tracer being retained for a longer period of time when injected into a borehole, the ascertained velocities are lower than they would have been if the tracer had been injected into natural fissures.

3 Bacteriological Indicators and Hydrological Conditions

Faecal contamination of springs, especially those that are used for human consumption, represents a serious epidemiologic problem (WHO 1997). The concentration of pathogenic organisms which originate from faecal contamination is usually low in waters, yet the variety of different pathogens is large; for example, bacteria such as *Campylobacter jejuni*, *Escherichia coli*, *Salmonella* spp., *Salmonella typhi*, *Shigella* spp. and *Vibrio cholerae*, viruses such as hepatitis A and parasites such as *Giardia lamblia* (Moe 1997). Generally, for water quality assessment not all potential pathogens are screened but only the indicator organism, or group of organisms, which frequently includes *E. coli* and coliform bacteria (WHO 1997). The term "coliform" is not based on taxonomic criteria but indicates lactose positive and oxidase negative genera of the family Enterobacteriaceae (Singleton and Sainsbury 2000). Faecal coliforms are present only in the gut of warm-blooded animals. Because the source of faecal coliforms is specific, as it results from human and animal excreta, it is thus a good indicator of fresh bacterial contamination. The main representative of faecal coliforms is *E. coli*. There is a sharp increase in the frequency of detection of *Salmonella* when the concentration of faecal coliform *E. coli* is greater than 200 CFU (colony-forming units) per 100 ml (Geldreich 1970).

Simultaneously with the tracer experiment at three selected springs (Boljunec, Osapska reka, Rižana) in the period from 12 November 2010 till 18 January 2011, water samples were screened for the presence of bacteriological indicators. Concentrations of *E. coli*, coliform bacteria and total aerobic bacteria were used to assess the quality of springs, to use the naturally present microbiota as an "indigenous tracer" and to observe the relation of microbiota with environmental variables.

3.1 Materials and Methods

For bacteriological analyses, RIDA®COUNT (R-Biopharm, Germany) test plates were used, the RIDA®-COUNT total plates for the enumeration of the total mesophilic aerobic bacteria and the differential RIDA®COUNT *E. coli*/Coliform plates for estimating the total coliform bacteria with differentiation for *E. coli* (Mulec et al. 2012a, b). Prior to starting the regular microbiological sampling campaign, along with the tracer experiment, the sensitivity of RIDA®-COUNT test plates for direct plating on the field was evaluated (Fig. 18). Shortly afterwards, the plates were on the field in triplicates aseptically inoculated with one millilitre of the water sample, subsequently transferred to a laboratory, and incubated for 24 h at 35 °C. The grown colonies were expressed as Colony-Forming-Units (CFU) per millilitre. For statistical analyses, the mean value was used.

Together with bacterial sampling, physicochemical measurements (temperature, pH, electrical conductivity-EC) were measured using a WTW Multiline P4 (Germany).

For statistical analyses, the following bacterial indicator groups were used: total bacterial counts (BAC), concentration of *E. coli* (ECO), the number of coliforms (COL), non-*E. coli* coliforms (NECCO) and non-coliform bacteria (NCOBA). NECCO was calculated as the number of *E. coli* colonies subtracted from the total coliform counts, and NCOBA represented the bacterial group which excludes coliform bacteria (Oarga et al. 2012). The sum of ECO, NECCO and NCOBA amounts to 100 % of all the bacteria detected by the used cultivation plates. Colonies, indicative of

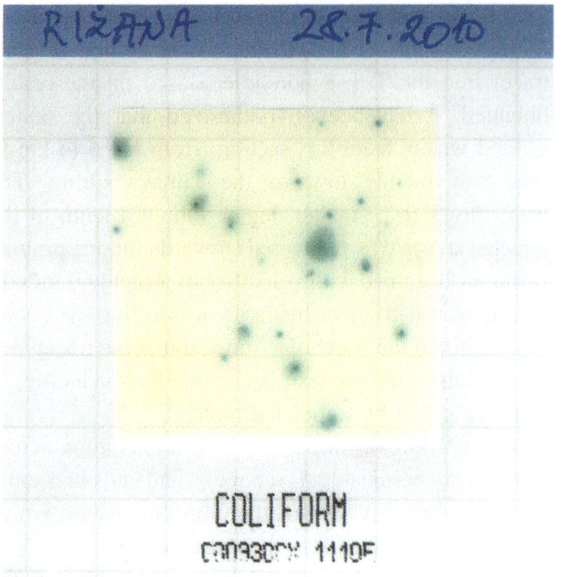

Fig. 18 Testing of the sensitivity of RIDA®COUNT Coliform plates (*blue* colony indicates β-galactosidase activity) for the direct field sampling at the Rižana spring (28 July 2010)

E. coli, were later not confirmed with an additional test. In downstream statistical analyses, the indicative values for *E. coli* were used. In the analyses with bacterial indicators, the data on the discharge of the springs and the concentration of the uranine tracer were applied as well (see Sect. 2). Analyses were carried out using PAST software (Hammer et al. 2001).

3.2 Results

During the research period, the temperature conditions were the most stable in the Rižana spring, the pH had the highest values in the Osapska reka spring, and EC was comparable among all the sites. Discharge was the highest in the Rižana spring, followed by the Osapska reka spring and the Boljunec spring (Pri pralnici). The uranine concentration was the highest in the Osapska reka spring, followed by the Boljunec spring and the Rižana spring (Table 2; see Sect. 2 for details).

In all the springs, the concentration of bacterial indicators fluctuated. In one millilitre of the samples, coliform bacteria were always present, while colonies, indicative of *E. coli*, were not (Fig. 19). However, this does not necessarily mean that in a given time the springs were *E. coli*-free but more likely that the

Table 3 Environmental parameters during the research period in the springs

Spring	Temperature (°C)	pH	EC (μS/cm)	Discharge (m^3/s)	Uranine (mg/m^3)
Boljunec	9.5–11.7	7.15–7.35	468–490	0.01–0.83	0.000–0.175
Osapska reka	7.6–11.1	7.91–8.28	457–509	0.08–10.55	0.000–0.511
Rižana	10.5–11.5	7.00–7.20	365–474	1.72–34.81	0.000–0.049

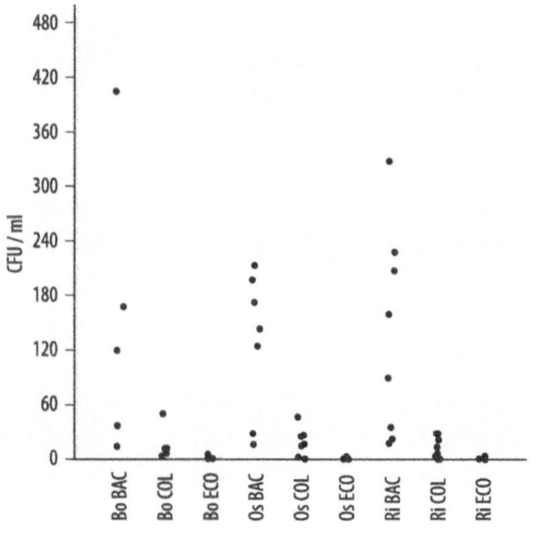

Fig. 19 Concentrations of bacteriological indicators (*BAC* total bacterial counts, *COL* coliform bacteria and *ECO E. coli*) during the winter of 2010/2011 in the springs (*Bo* Boljunec, *Os* Osapska reka, *Ri* Rižana)

concentration of this bacterium was below the detection limit of this method. The highest portion of *E. coli* in the community based on the RIDA®COUNT cultivation tests was detected in the Osapska reka spring (2.93 %), while in the other two springs *E. coli* represented a smaller portion of bacterial communities, up to 1.67 % in the Boljunec spring and a maximum of 1.43 % in the Rižana spring. The percentage of the NECCO group varied; the highest percentage was observed in the Boljunec spring (31.30 %), followed by the Osapska reka spring (22.00 %) and the Rižana spring (18.52 %). The rest of the bacterial biomass was attributed to NCOBA; in the Boljunec spring, it ranged from 68.70 to 95.83 %, in the Osapska reka spring from 78.00 to 96.55%, and in the Rižana spring from 81.48 to 97.14 % (Table 4).

Pearson's correlation coefficient (r) was calculated to observe the potential linear relationship between cultivable bacterial groups and physicochemical

Table 4 The bacterial community structure based on the cultivation on RIDA®COUNT test plates

Spring	ECO (%)	NECCO (%)	NCOBA (%)
Boljunec	0.00–1.67	3.57–31.30	68.70–95.83
Osapska reka	0.00–2.93	3.45–22.00	78.00–96.55
Rižana	0.00–1.43	1.43–18.52	81.48–97.14

(*ECO E. coli*, *NECCO* non-*E. coli* coliforms and *NCOBA* non-coliform bacteria)

parameters. The temperature against the bacterial groups (BAC, COL, ECO, NCOBA and NECCO) showed a strong positive correlation in the Boljunec and Rižana springs but a strong negative correlation ($r < -0.70$) in the Osapska reka spring. The results for the springs Boljunec and Osapska reka should be re-confirmed, due to the low number of available physicochemical measurements (Table 5).

Relations between the pH and microbial groups showed a negative correlation. EC showed a strong positive correlation in the Rižana spring with bacterial indicators; statistical significance was observed in the case of *E. coli* ($p = 0.043$). The correlation results for the Boljunec and Osapska reka springs are merely informative, due to the low number of the available pH measurements (Table 5).

Positive correlations were observed between the discharge and bacterial concentrations. In the case of the Rižana spring, all the tested bacterial groups showed very strong positive ($r > 0.70$) and statistically significant correlations ($p < 0.05$). The discharge had a statistically significant correlation also with BAC ($p = 0.009$) and NCOBA ($p = 0.005$) in the Boljunec spring (Fig. 20). The discharge proved to be one of the most important parameters influencing bacterial concentrations in these two springs, but not in the Osapska reka spring. This might be connected with the fact that in contrast to the Boljunec spring and the Rižana spring, the Osapska reka spring is not active all year round.

In the analysis, we also tested the correlations between the concentration of uranine and bacterial

Table 5 Pearson's correlation coefficient (r/p, in bold $p < 0.05$) between cultivable bacterial groups and physicochemical parameters

Indicator group	Temperature	pH	EC	Discharge	Uranine
Boljunec (n = 5)					
BAC	0.988/0.099*	−0.814/0.395*	−0.776/0.435*	0.961/**0.009**	0.924/0.076
COL	0.755/0.455*	−0.080/0.949*	−0.017/0.989*	0.777/0.122	0.989/**0.011**
ECO	0.933/0.234*	−0.918/0.260*	−0.891/0.300*	0.865/0.058	0.981/**0.019**
NCOBA	0.975/0.144*	−0.852/0.350*	−0.818/0.390*	0.974/**0.005**	0.902/0.098
NECCO	0.578/0.608*	0.159/0.898*	0.221/0.858*	0.758/0.137	0.985/**0.015**
Osapska reka (n = 7)					
BAC	−0.977/0.137*	−0.995/0.065*	−0.990/0.090*	0.430/0.336	0.114/0.809
COL	−0.993/0.078*	−1.000/**0.006***	−0.999/**0.031***	0.070/0.882	−0.063/0.894
ECO				0.701/0.079	0.184/0.693
NCOBA	−0.970/0.155*	−0.991/0.083*	−0.986/0.108*	0.508/0.244	0.154/0.741
NECCO	−0.993/0.078*	−1.000/**0.006***	−0.999/**0.031***	0.011/0.982	−0.077/0.869
Rižana (n = 8)					
BAC	0.666/0.071	−0.339/0.412	0.450/0.264	0.936/**0.001**	0.549/0.158
COL	0.641/0.086	−0.176/0.677	0.569/0.141	0.849/**0.008**	0.383/0.350
ECO	0.829/**0.011**	−0.153/0.718	0.723/**0.043**	0.824/**0.012**	0.028/0.948
NCOBA	0.657/0.077	−0.352/0.393	0.428/0.290	0.930/**0.001**	0.559/0.150
NECCO	0.566/0.144	−0.166/0.695	0.503/0.204	0.787/**0.020**	0.403/0.322

(*BAC* total bacterial counts, *COL* coliforms, *ECO E. coli*, *NCOBA* non-coliform bacteria and *NECCO* non-*E. coli* coliforms)
* Correlation analysis based on only three measurements of temperature, pH and EC

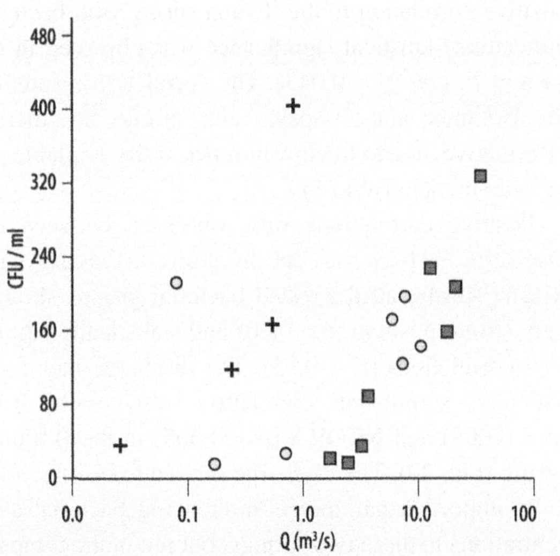

Fig. 20 Relation between the concentration of the total cultivable bacteria (CFU/ml) and the discharge in the springs (+ Boljunec, *bullet* Osapska reka, *square* Rižana)

groups. With pulses of the tracer, bacterial concentration was positively correlated ($p < 0.05$) for the majority of groups in the Boljunec spring (COL, ECO and NECCO) and in the Rižana spring, whereas the Osapska reka spring showed no or negligible relationship ($r < 0.184$; Table 5). Again, this might be attributed to the different nature of the Osapska reka spring.

The nonparametric Kolmogorov–Smirnov test was used for testing whether the concentrations of BAC, ECO and COL among all three springs have equal distribution. The permutation p value (10,000 permutations) was >0.05 for all samples; this resulted in the adoption of the null hypothesis that the samples (BAC, COL and ECO) came from a population with equal distribution. This indicates a similar situation in the recharge area for all three springs with regard to the biological load.

The Boljunec, Osapska reka and Rižana springs occasionally had a high concentration of total cultivable bacteria, and in all the springs isolates indicative of *E. coli* were retrieved. A similar distribution of bacterial indicators in all the karst springs may indicate a comparable situation in the recharge area with regard to the biological load. In the given hydrological situation (winter 2010/2011), the discharge was correlated with bacterial indicator groups and can be thus considered an important factor for the permanent karst springs, Boljunec and Rižana, but not for the periodic Osapska reka spring. Other correlations between physicochemical and bacterial indicators can be attributed to the specific conditions at each spring.

The construction of railway tunnels through the recharge area of the Boljunec, Osapska reka and Rižana springs might affect hydrological conditions of the springs; it is therefore advisable to continue bacteriological monitoring during the construction and after the completion of the works. In any case, the bacteriological load of the groundwater in this area should not be increased further. Future monitoring activities can be based on the presented methodology because it is simple and low-cost and can be upgraded to a more advanced one.

4 Risk of Contamination of Karst Water in the Impact Area of the Planned Railway Line

4.1 Risk Assessment

A review of the results of the conducted tracer tests has shown good permeability of karst and a concentrated and fast flow through the karst aquifer. This is especially true of the northern part of the area in question with a highly developed karst drainage network between the ponor of the Reka River and the springs along the Gulf of Trieste in Italy. If a potential contamination from the area of the route reached the underground flow of the Reka River through the vadose zone, the further transfer to the Timava springs would be very fast; during the highest discharges, the velocities of around 200 m/h can be expected, and of 40 m/h during the lowest discharges. The tracing from the surface near Sežana during a relatively low discharge of the Timava spring ascertained velocities of around 40 m/h. It has therefore been deduced that from this area (from 0 to 6 km), the substances dissolved in the water would appear in the Timava spring during a low water level about a month later; but during a very high water level they could appear within one week. Since the share of the Reka River in the Timava spring has been estimated at around merely 30 % (Cività et al. 1995), any exceeding of the allowed concentrations could be detected in the Timava spring only after greater contamination (accidents) in the recharge area, in accordance with the flow of water under the given hydrological conditions. It should be noted that the most dangerous substances are those that cause harmful effects even in very low concentrations (e.g. oil derivatives).

A similar estimate also applies to the section between 6 and 9 km, only that there the tracing in the T 1–8 borehole also proved possibility of a smaller portion (10 %) flowing out towards the Boljunec spring in Italy (the Pri pralnici permanent spring and the Jama intermittent spring). During a high water level, the potential contamination in this spring would appear after approximately 4 days. The third spring in Boljunec (Na placu) is not directly affected by the planned railway line.

Waters flow from the flysch area in the northern part of the T 2 tunnel towards the contact with karst rocks in the area of the Beka-Ocizla cave system; from there a direct connection to the Pri pralnici spring in Boljunec has been proved (during a low water level, the apparent flow velocity is 30–40 m/h); the substances dissolved in the water from the area of the railway route could appear there within approximately 4 days during a low water level or even earlier during a high water level. The possibility of a slower flow towards the Rižana spring was also confirmed. The portion of the tracer recovered in the Rižana spring was small, but because it is an important water source, this underground water connection must be taken into account when planning protective measures.

For the area between 12 and 13.5 km, the flow towards the springs Pri pralnici in Boljunec, Osapska reka and Rižana can be assumed; yet, the ratio between them cannot be determined more precisely.

In the southern part of the route between 13.5 km and the southern portal of the T 2 tunnel, the waters flow out mainly towards the Rižana and Osapska reka springs. During a high water level, the tracer appeared in the Osapska reka spring after three days, and about a day later in the Rižana spring. Under such conditions, we could predict the risk of the contamination of the Rižana by observing Osapska reka. When the water level drops, the portion of the flow towards the Rižana spring probably increases. Potential contamination from this section of the route might also appear in the Pri pralnici spring in Boljunec to a small degree (a few percent) and later.

Despite the fact that during the tracer test near Črnotiče we injected the tracer on the surface, it has been demonstrated that the flow velocities in the vadose zone were relatively high through the more permeable fissures. Our comprehensive research into the karst of Postojna has shown that this flow in the case of spills of a greater quantity of water-soluble substances, can reach relatively high velocities, even up to 80 m/h, and consequently a fast transfer of a part of the contamination to the continuous underground flows. However, a greater part of the contamination is retained in the poorly permeable part of the vadose zone for several years or decades and is rinsed out towards the springs with a longer time delay, depending on the precipitation conditions (Kogovšek 1997, 2000, 2010). Such events may occur when dangerous substances are spilt during traffic accidents, especially if the substances are not soluble in water, as these are transferred through the karst aquifers much slower than soluble substances, which means that they accumulate. During rainy periods, the transfer of contamination from the surface is immediate and continuous (Kogovšek 2000), whereas in the dry periods it is minimal or even non-existent. Hence, in the case of the constant, smaller contamination that will appear during the operation of the railway line, the contaminants will be rinsed away each time it rains, which indicates steadier, yet constant contamination.

4.2 Protection of Important Karst Water Sources in the Impact Area of the Route

From the aspect of the supply of drinking water, the Rižana spring, which has been captured for the water supply of the coastal municipalities, is the most important spring in the area in question. This area needs large quantities of drinking water, especially in the summer when the discharges of the Rižana spring are too small, and consequently water must be provided from additional sources. In 2008, the Decree on determining the drinking water protection area for the aquifers of Rižana was published in the Official Gazette of the Republic of Slovenia, which defines three protection zones. In one section (between 13 and 16 km), the planned route runs across the outer water protection zone (III), within which the construction of the railway line is permitted if it is in accordance with the national or municipal site plan, adopted pursuant to the spatial planning regulations, if a comprehensive assessment of the impact on the environment has been performed for this site plan in accordance with the regulations governing the comprehensive assessment of the impact on the environment, and if protective measures are implemented due to the impact of construction on the water regime and status of the water body which show, based on the results of the risk assessment, that the risk of contamination from construction is acceptable.

The described tracer tests have shown that the waters from the southern part of the planned route flow underground towards the Rižana spring; therefore, a negative impact on the quality of this water source is possible during the construction and in the case of potential contamination during the operation. The Osapska reka spring is an intermittent spring which is active only during high water levels and has not been captured for the drinking water supply. The existing data show that this spring is also located within the impact area of the planned route.

The springs in Boljunec on the Italian side have not been captured for water supply, while the Pri pralnici spring is being used for fish farming. The established good connection with the Beka-Ocizla cave system points to the high endangerment level of this spring and the possibility of a direct impact of the planned construction. In light of the established characteristics, this area should be classified under the inner water protection zone with a strict regime.

The northern part of the planned route runs across the southeast edge of the aquifer of the Classical Karst. The groundwater from this area flows towards the northwest and the Timava spring along the Gulf of Trieste. Speleological research has ascertained the existence of larger and well-permeable karst channels between Škocjanske jame cave and the Timava spring (on the Slovenian and Italian sides several shafts have been explored in which the underground flow of the sinking Reka River can be accessed from the surface). The high velocities of the underground flow in this area have been confirmed with tracer tests. Therefore, the activities affecting the karst south of Divača may influence the quality of the Timava spring; however, due to the relatively small portion of the Reka River in the recharging of the spring, only contamination with dangerous substances presents a greater danger.

In the area of the Classical Karst, the station for pumping underground karst water in Klariči, from which the inhabitants of five karst municipalities are supplied, is of exceptional importance to the water supply. The area where the water is captured lies north of the main underground flow of the Reka River between the Škocjanske jame cave and the Timava spring, which is why the possibilities of a negative impact of the railway construction are slight.

In light of the described potential endangerment of the above-mentioned water sources, a comprehensive analysis of the quality of the springs should be performed prior to commencing construction and consequently the so-called zero state should be determined (monitoring the quantity and quality under various hydrological conditions). More detailed monitoring, particularly of the springs Rižana, Pri pralnici in Boljunec and Osapska reka, during the construction and afterwards, during the operation, would show the actual impact of the planned activity. When planning this activity, one must take into account the special characteristics of karst aquifers which, on the one hand, enable a very fast water flow and a transfer of substances across the more permeable karst channels and fissures; on the other hand, during dry summer conditions, the contaminants may accumulate in the vadose zone and be rinsed to the springs only after the first intense and abundant precipitation, usually in autumn. It would therefore be sensible to monitor select flood pulses in which the dynamics of the sampling would be suitably adapted to the precipitation and hydrological conditions.

Caves and Hydrology of the Contact Karst of Beka and Ocizla

The Beka-Ocizla cave system extends at the contact between Palaeocene limestone and Eocene flysch at an altitude of 350 m in a shallow depression called Loke. Layers of limestone generally dip towards the north-east; the most distinct fault structures are located in the Dinaric (NW–SE) and cross-Dinaric (NE–SW) direction. Three larger and one smaller allogenic streams flow in from flysch to the contact with limestone and sink into limestone.

According to the basic geological map 1:100,000, Trieste sheet (Pleničar et al. 1969), it is the contact between Palaeocene/Eocene limestone and Eocene flysch. The area belongs to the morphostructural unit of Podgorski kras (Bosák et al. 1999, 2004), which has typical narrow stripes of flysch between limestone, as the territory structurally belongs to the imbricate structure of Čičarija (Pleničar et al. 1973; Placer 1981, 2005; Rižnar et al. 2007; Placer et al. 2010). There is no evidence of younger marine deposition than Eocene in the southwestern part of Slovenia (Zupan Hajna et al. 2008). Younger sediments occur only in caves and very rarely on karst surface (different soils and few remains of terrigenous sediments). Calibrated data from paleomagnetic research of cave sediments of the area contributed to reconstruction of speleogenesis, deposition in caves, and indirectly to evolution of karst surfaces and succession of tectonic movements. These data indicate that evolution of caves in southwestern Slovenia took part within one post-Eocene karstification period (Zupan Hajna et al. 2010).

Six known caves are connected to the Beka-Ocizla cave system (Fig. 1): Ocizeljska jama (Ocizla cave), Blažev spodmol (Blaž's rock shelter), Maletova jama s slapom (Maletova jama cave with waterfall), Jama z naravnim mostom (Cave with the natural bridge), Jurjeva jama v Lokah (Jurjeva jama cave in Loke) and S-4/Socerb. All the caves are interconnected with human-size passages, except for the passages of the S-4/Socerb, which, according to ground maps in the Cave Registry, come very close to the cave Jama z naravnim mostom. The cave system has been explored extensively; at the end of the nineteenth and at the beginning of the twentieth century the cave Ocizeljska jama was explored by members of the Trieste S.A.G. (Società Alpina delle Giulie) who measured the cave all the way to Staro dno (Bratoš and Sancin 1984). Later, the cave was visited by members of Caving Club Železničar; intensive exploration was then continued by members of the JOSPD Trst Speleological Society, headed by Stojan Sancin. The members of this society submitted a vast number of records to the Cave Registry, either on their explorations or their survey of the cave system and are likewise the authors of the majority of the cave maps. For the needs of geological mapping, Ocizeljska jama, a part of the Jama z naravnim mostom, Jurjeva jama and the S4/Socerb were remapped in 2001 and 2009 by the Karst Research Institute ZRC SAZU.

The maximal distance between all six entrances is about 500 m. The cavernous porosity (i.e. cavernosity) of the territory is high, as the total length of the currently known passages is about 4,422 m. The greatest depth in the cave system, reached so far, is 157 m below the entrance to Ocizeljska jama.

M. Knez et al. (eds.), *The Beka-Ocizla Cave System*, Cave and Karst Systems of the World,
DOI: 10.1007/978-3-319-04456-9_5, © Springer International Publishing Switzerland 2015

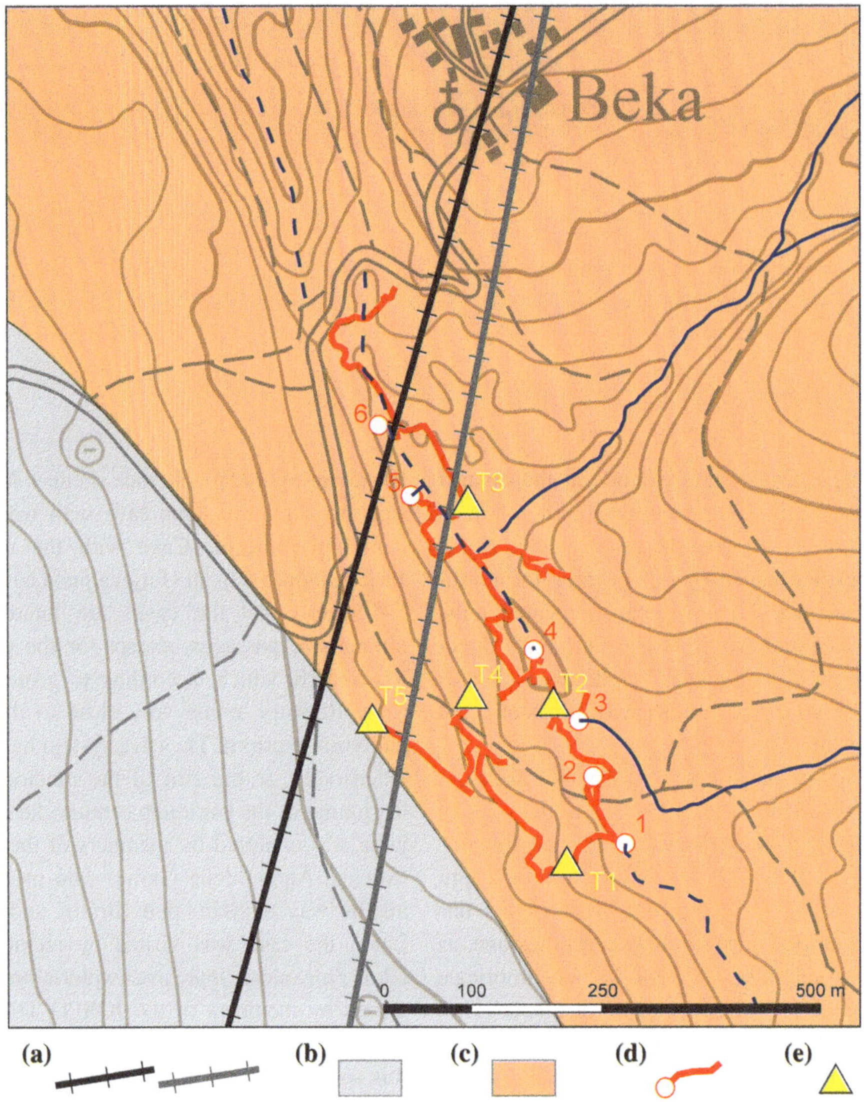

Fig. 1 Location of the caves in the Loke depression (according to the Cave Registry). *1* Ocizeljska jama, *2* Blažev spodmol, *3* Maletova jama s slapom, *4* Jama z naravnim mostom, *5* Jurjeva jama and *6* S-4/Socerb; **a** route of the tunnel (*grey*—proposed, *black*—accepted), **b** limestone, **c** flysch and transitional beds, **d** cave entrance and underground passages, **e** hydrological observation stations

Entrances to the caves were linked to the national geodetic network by theodolite survey. The connection was set up across a trigonometrical point of the IV order, No. 87, southwest of Beka and the trigonometer on Slavnik Mountain. Polygonal points at the entrances to the caves were stabilized with brass plugs Φ 6 cm and marked "IZRK Postojna, izmera jame" ("IZRK Postojna, cave measurement"). Five polygonal points were stabilized, the coordinates of which in Gauss-Krüger coordinate system are shown in Table 1.

1 Geological and Speleomorphological Characteristics of the Caves

Entrances to the caves of the Beka-Ocizla cave system are mostly located in the central part of the depression. The lowest is the entrance to Jama z naravnim mostom, i.e. 346 m above sea level. The highest is the entrance to the S4/Socerb cave at the altitude of

Table 1 Coordinates of the stabilized polygonal points at the cave entrances

Cave	y	x	z (m)
Ocizeljska jama, collapse doline (bottom)	414,221.04	50,303.24	322.42
Ocizeljska jama, collapse doline (edge)	414,229.36	50,337.14	356.13
Maletova jama s slapom	414,177.10	50,448.73	352.20
Jama z naravnim mostom	414,124.65	50,497.29	346.18
Jurjeva jama	413,974.12	50,698.09	360.71

371 m, located near the pass towards the gorge Griž-nik. The biggest entrance to the system is presented by the collapse doline Ocizla, which is 34 m deep and measures 50 m in diameter.

Almost all the mentioned cave entrances function as periodic stream sinks, except in the case of S4/Socerb and the entrance to Blažev spodmol. The most permanent inflow of water is the one to Maletova jama s slapom and Jama z naravnim mostom; the inflow to Jama z naravnim mostom is the last to dry up. The streams sinking into Ocizeljska jama and Jurjeva jama are intermittent, active only during a higher amount of precipitation.

In the explored area (Zupan Hajna 2004) between the outermost entrances to the cave system, i.e. between the caves Ocizeljska jama and S4/Socerb, the Alveolina-Nummulites limestones of the Lower Eocene age gradually pass into flysch rocks, across the so-called transitional beds (Jurkovšek et al. 1996). The transitional beds consist of marly limestone and marl and are located at the bottom of the depression; the entrance parts to all the caves of the system are located there. A gorge of the stream which sinks into Jama z naravnim mostom has been formed into marly limestone. The morphology of the contact between limestone and flysch in the Loke depression and the geological cross-section of the contact are shown in Fig. 2. The cave system developed at the contact between limestone and transitional beds; of special interest are the entrances to the caves Jurjeva jama and to S4/Socerb which are in the slaty marls. The beds in the investigated territory generally dip towards the northeast (70/20–40).

2 The Caves in the Beka-Ocizla System

2.1 Ocizeljska Jama

Data from the Cave Registry: Reg. No.: 1,003; synonym: Ocizla; entrance location: $y = 414{,}229$, $x = 50{,}337$, $z = 356$ m; length 2,780 m, depth 157 m.

Renewed measurements of the Ocizeljska jama (Gabrovšek et al. 2001) were conducted within the scope of geological and speleological research and repeated in 2009. The difference in altitude between the entrance (356 m) and the lowest point in the cave (223 m) was 133 m, while the sump at the end of the cave extends even 24 m deeper. Oscillation of water table in the cave was quite high. The new measurements of the water level in the cave are described and shown in Sect. 4.

The Ocizeljska jama is the biggest and the longest cave of the Beka-Ocizla cave system (Figs. 3, 4, 5, 6, 7). It is accessible through a collapse doline, 34 m deep and 50 m wide, and through the cave Blažev spodmol. Below the collapse doline, a meander and two shafts lead to an extended network of horizontal passages that developed at the horizon extending to a depth of 80 and 90 m (the passage Novi rov and the chamber Peterokraka dvorana). A steep passage with two vertical drops (Meander) connects Peterokraka dvorana and the lowest horizontal level, the passage Rov velike razpoke. A steep, up to 30 m high, meander-like passage between Peterokraka dvorana and Rov velike razpoke follows the dip of the tectonic plane (220/40–50) and has developed beneath it (Fig. 8). The

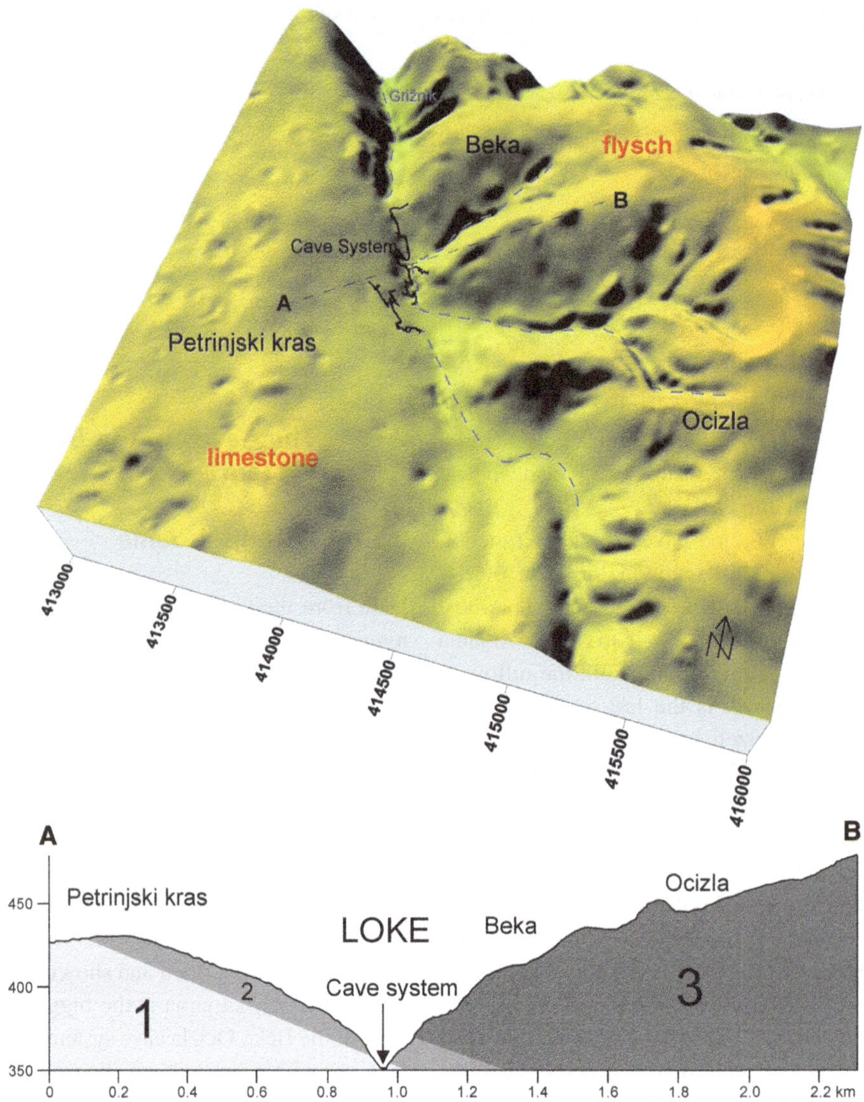

Fig. 2 Morphology of the contact between limestone and flysch in the Loke depression and the geological cross-section of the lithological contact. The contact between *1* Lower Eocene Alveolina-Nummulites limestones and *3* Eocene flysch (alternation of marls, silicate sandstones and breccias; 3) is gradual over so-called *2* transitional beds (permeable slaty marls and marly limestone)

parallel tectonic structure is responsible for formation of the passage Rov velike razpoke which is over 40 m high.

The beds of marly limestone and limestone on the surface are directed towards the northeast with a dip of 70/20–30. The same dip of the dark Alveolina-Nummulites limestone beds can be found throughout the cave. A smaller part of the cave passages has been formed along the bedding planes, particularly the bottoms of the stepped shafts. The formation of the cave and the orientation of the cave passages were mostly influenced by the system of parallel tectonic fault planes with dip 220/40–50.

The final part of the cave, the passage Rov velike razpoke, is entirely formed along the before mentioned fault plane with dip 220/50 and has a look of a high and narrow opened fissure (Fig. 9). Here the mentioned tectonic plane represents the upper wall of the passage and can be followed for an additional 40 m up the inclined chimney. An open part of the passage Novi rov has been formed along the same fault plane which can be seen at several places. The passage along it has

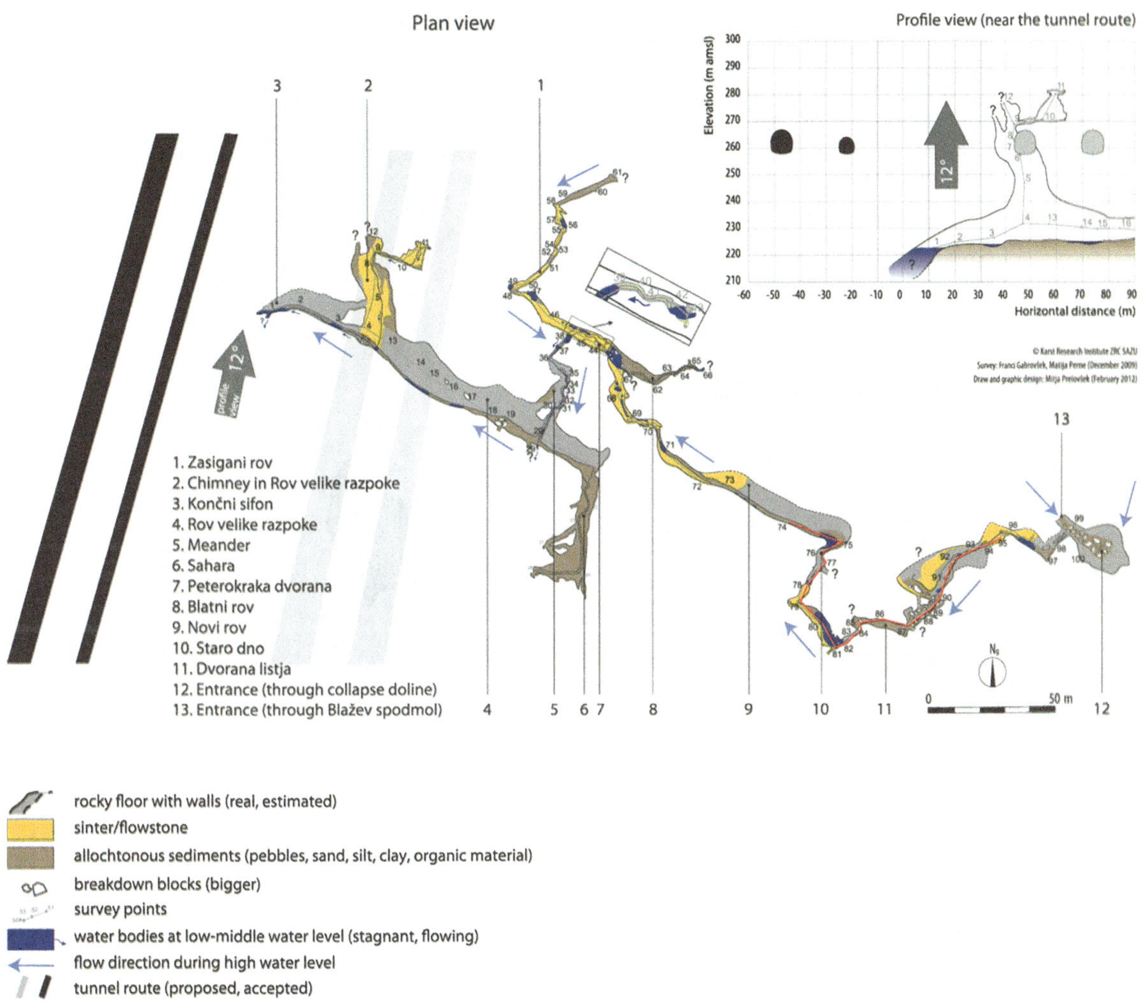

Fig. 3 The plan of the cave Ocizeljska jama with the position of proposed (*grey*) and accepted (*black*) tunnel route. The movement of the tunnel route is due to collision of the tunnel with the cave passage

either been widened by a few metres by dissolution or the same fault plane cuts through the same passage, which is not widened but located next to the clay of the inner fault zone. The same fault plane can also be found in the upper part of the cave (Fig. 10), i.e. at the top of the first series of stepped shafts at depth between 35 and 70 m. Parallel to it, there are the first parts of the passages Zasigani rov and Blatni rov at depth of around 80 m. In the second series of stepped meander, at depth between 90 and 120 m, we can observe parallel fault planes with the same dip a few more times.

Thus the stepped meander has formed along the above-mentioned tectonic planes and cut down into three stairs. In all the cases the ceiling of the passage is formed by the fault plane itself. When the passage lowers and nears the ceiling, i.e. the fault plane, a network (labyrinth) of smaller passages of phreatic shape was formed to the left and right of the plane; there the cave continues along these passages, parallel to the fault plane. Rock features demonstrate that the passages along the fault planes were initiated by dissolution and have later been extensively widened by mechanical erosion as well. A good example is the passage Rov velike razpoke, leading to the terminal sump in Ocizeljska jama.

The main fault plane was mechanically weakened at several places by transverse fissures; through those fissures water was able to break through and at those

Fig. 4 A passage in the cave

places the cave continued transversely to the fault plane by smaller passages. When the water/passage reached the next parallel fault plane with the same dip as the main fault plain (220/40), the process of formation of the transverse passages can be repeated.

At the bottom part of the cave a system of fissures with dip 320–310/70–90 can be also observed; the fissures are followed by individual segments of the cave passages. In the entrance collapse doline and in the passage Novi rov few fissures with N–S direction can be observed, but they did not have a direct impact on the formation and distribution of the existing cave passages.

2.2 Blažev Spodmol

Data from the Cave Registry: Reg. No.: 1,004; synonym: Bečka jama; entrance location: y = 414,229, x = 50,337, z = 356 m; length 120 m, depth 30 m.

The entrance to the cave Blažev spodmol is through a small collapse doline. Its main passage ends in the big collapse doline of the Ocizeljska jama. Another, longer passage directed northwards connects Blažev spodmol with the cave Jama z naravnim mostom.

The upper part of the entrance collapse doline is formed in marly limestones, while the passages are formed in the Alveolina-Nummulites limestones. The slope of the collapse doline is made of pebbles, sand and clay originated from flysch rocks. The big scallops on the walls in the bottom of the main passage indicate a slow water flow that once flowed through the passage. The passage is now partially filled with sediments which, according to their locations, once filled up the entire passage. There are also many remains of disintegrated and eroded older flowstone.

2.3 Maletova Jama s Slapom

Data from the Cave Registry: Reg. No.: 729; synonyms: Maletova jama, Korošica na hribu, Jama s slapom; entrance location: y = 414,177, x = 50,448, z = 352 m; length 78 m, depth 26 m.

At the entrance, a periodic stream falls over marly limestones in a waterfall 8.5 m deep. Opposite to the waterfall and at the entrance to the cave, a trench is located. It is several metres long; the bottom elevates towards the southwest and is covered with gravel. According to its shape and location, it might not be

Fig. 5 Surveying in Ocizeljska jama

tectonic in origin since the entrance passage is not related to any tectonic structure, but is perhaps a remnant of a former cave passage that followed dip of strata, filled up with sediment, and the ceiling of which has been denuded. The cave continues beneath the waterfall towards the west and then turns towards the

Fig. 6 Omega-shaped channel in Blažev spodmol

north in its final section. The cave is about 80 m long and descends for 26 m, i.e. to an altitude of 326 m. A dye tracing test with the injection point at Maletova jama s slapom was done in 2001 (Gabrovšek et al. 2001); the tracing test confirmed that the water flowed towards the Boljunec spring (Kogovšek and Petrič 2003).

Fig. 7 Climbing up the Retroverzni rov (entrance series) in Ocizeljska pečina

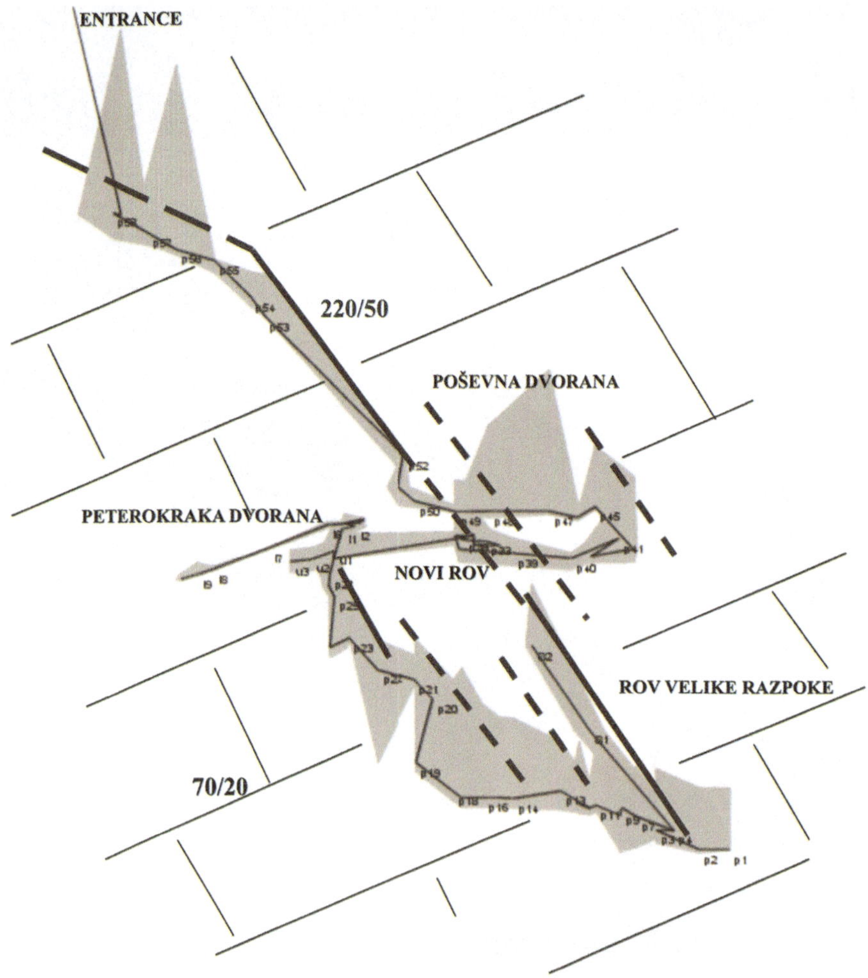

Fig. 8 The cave Ocizeljska jama (profile 130°). A single fault plane (dip 220/50) can be traced throughout the cave. The system of parallel fault planes was an important factor of the entire cave system development, orientation of the main passages and direction of the water flow (measurements and the cave plan Karst Research Institute ZRC SAZU 2001)

2.4 Jama z Naravnim Mostom

Data from the Cave Registry: Reg. No.: 723; synonyms: Miškotova jama v Lokah, Miškotova jama z naravnim mostom; entrance location: y = 414,124, x = 50,497, z = 346 m; length 447 m, depth 79 m.

Water flows into the cave Jama z naravnim mostom through a 100 m long gorge formed in marly limestones, from the narrow valley beneath the Beka village. A natural bridge is located at the entrance and crosses the gorge where it reaches a collapse doline in front of the cave entrance. The top of the entrance passages is formed in marly limestones gradually passing into foraminiferal limestones. The beds dip towards the northeast (70/20). The main segments of the passages follow the faults in the Dinaric (NW–SE) direction; the faults dip slightly towards the southwest or they are vertical. In the cave, fissures with dip 310/70–90 can also be found. Water flows into the cave along a passage which is 2 m wide and up to 8 m high. The new survey of the entrance passages confirmed the correctness of the lengths and directions of the passages at the plan made by cavers; hence, the inner parts of the cave have not been remapped. About 200 m from the entrance, the main meander-like passage turns into a periodic siphon. In a side passage behind, a shaft Vodnjak was found with the depth estimated to about 100 m. The shaft has not been surveyed yet, but its bottom is possibly below the level of the sump in the cave Ocizeljska jama. About 80 m further a connecting passage from the cave Jurjeva

Fig. 9 The passage Rov velike razpoke at the *bottom* of the cave Ocizeljska jama is 100 m long and over 40 m high; it has formed along a fault plane with dip 220/50

Fig. 10 The first series of stepped shafts below the collapse doline has been cut beneath the fault plane (the passage ceiling). The figure shows the fault plane where it changes its inclination

jama enters the main passage. From there the cave continues at more or less the same altitude as the former final sump, named Končni sifon, which recently has been surpassed. Behind this former sump, the main passage continues at the same altitude for another 150 m and then descends in a series of shafts for 50 m. At the bottom of the stepped shafts, at approximately 267 m above sea level, the cave ends in a smaller chamber with a sump of unknown depth.

2.5 Jurjeva Jama v Lokah

Data from the Cave Registry: Reg. No.: 636; synonym: Brezno v Beki; entrance location: $y = 413{,}974$, $x = 50{,}698$, $z = 360$ m; length 65 m, depth 41 m.

The entrance to the cave Jurjeva jama is a sink of periodic stream which flows along the narrow valley southwest of the Beka village. The opening of the entrance shaft is partitioned by a flowstone natural bridge (Fig. 11). The entrance shaft is 23 m deep; its bottom, around 5 m wide, is filled with gravel. It is elongated along a fissure with dip 310/90. The upper part of the entrance shaft is made of slaty marl and marly limestone (dip 70/40) that gradually pass into foraminiferal limestones at the bottom of the shaft. The cave continues through a narrow section in the northern wall of the passage where an 8 m long narrow meander follows and ends in a shaft of 10 m. There is a smaller chamber below, 5 m in diameter. The meander that leads to the lower parts of the cave is open along a tectonic plane with dip 220/50; the chamber beneath the 10 m shaft has been formed along the fault with dip 220/80. After the small chamber the cave continues in a 20 m long low passage that leads to the cave Jama z naravnim mostom.

2.6 S-4/Socerb

Data from the Cave Registry: Reg. No.: 5,772; entrance location: $y = 413{,}943$, $x = 50{,}783$, $z = 371$ m; length 90 m, depth 63 m.

The entrance to the cave S-4/Socerb was artificially enlarged to a blowhole 40 cm in diameter. It is located on the slope, a few metres above the usually dry bed of the stream that sinks into the cave Jurjeva jama. The cave profile is shown in Fig. 12. The entrance shaft is 3 m deep; its bottom opens up into a smaller chamber. The entrance and the entrance chamber are formed in marly limestone with dip 70/40, which alternates with

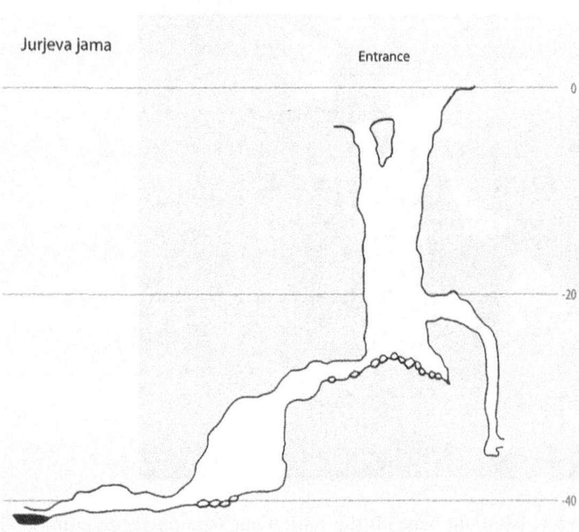

Fig. 11 Extended elevation of the cave Jurjeva jama (Karst Research Institute ZRC SAZU 2001)

Fig. 12 Extended elevation of the cave S-4/Socerb (Karst Research Institute ZRC SAZU 2001)

slaty marl in the top part and along the fault with dip 210/80. After a descent of about 20 m through narrow passages to the collapse chamber, the passage turns into a 47 m deep shaft. The shaft is formed along the fault with dip 210/80. The bottom of the shaft, at 308 m above sea level, is 5 m wide and covered with flysch loam below which the continuation will perhaps be possible. In the chamber, beneath the first narrow passage, marly limestone beds dip towards the north-east (70/40). About 18 m above the bottom of this shaft there is a 13 m long horizontal passage through which a periodic small stream flows in, falls to the bottom of the shaft in a waterfall and disappears in clay. This water probably flows towards the cave Jama z naravnim mostom which is, according to measurements, only a few metres away. The chamber behind the second narrow passage and the collapse chamber in front of the entrance to the shaft are formed underneath a single tectonic plane with dip 100/50.

3 Hydrogeological Characteristics

Three larger streams and a smaller one flow into the depression from flysch and sink into marly limestone. Together, they drain the flysch surface of approximately 3.5 km^2 (Mihevc 1991). The surface waters disappear into the karst at the contact which is, on average, located at 350 m above sea level. Some of the cave entrances function as periodic sinks of the surface waters that flow in from beneath the villages of Beka, Ocizla and Petrinje.

The biggest sinking stream initiates beneath Petrinje and flows into the cave Ocizeljska jama, but only during very high waters. Usually, it already disappears in its carbonate river bed. The stream that flows along the gorge north of the village of Ocizla sinks into the cave Maletova jama s slapom as soon as it flows off the flysch to the limestone. A parallel stream which flows along the ravine south of the Beka village sinks into the cave Jama z naravnim mostom. This stream continues to flow across the surface for a while even after it flows off the flysch to the marly limestones but it continuously loses water on the way to the cave entrance. A small gorge has cut into the transitional beds, while tufa is precipitated along the bottom of the river bed whenever water flow encounters smaller barriers in them. A smaller stream flows southwest of the village Beka only during intensive precipitation and sinks into the cave Jurjeva jama.

The water from the entire cave system flows towards the Boljunec springs (Sancin 1988). Its source is located at an altitude of 70 m above the Gulf of Trieste at the Adriatic Sea. This connection was confirmed by the dye tracing test in 2001 done by the Karst Research Institute ZRC SAZU (Gabrovšek et al. 2001; Kogovšek and Petrič 2003) when uranine was injected into the cave Maletova jama s slapom.

4 Measurements and Analysis of Hydrological Parameters in the Beka-Ocizla Cave System

During flood events the flow in the Beka-Ocizla cave system is very complex; water enters the system at several locations, the caves are hydraulically connected, directions of the flows are not unambiguous, the numerous constrictions present hydraulic bottlenecks that cause back flooding. A detailed study of the water conditions in the cave would require the set-up of a large number of observation points and measurements of the flow rates of all inflowing allogenic water. Still, one would probably not be able to avoid a deficient characterization, as the cave system most likely extends far beyond what has been explored so far.

Hydrological observations conducted in the cave include the mapping of the traces of past flood events and the continuous monitoring of the water level and temperature at five different locations in the system. The longest record has been obtained at the deepest point, the final sump in Ocizeljska jama, which is also the closest location to the envisaged route of the tunnel.

All observation places were equipped with autonomous loggers (Eijkelkamp Diver^{TM} and Hobo U 24), which record pressure and temperature at arbitrary time intervals. If not stated otherwise, 10 min intervals between subsequent measurements were selected. The position of observation places is marked in Fig. 1. Hereon, we give a short description of these locations.

T 1 Dvorana listja (Hall of Leaves) in Ocizeljska jama The measurement point is below the entrance shaft series (Figs. 1 and 13). The point was selected due to the deposits in the entrance collapse doline after the major flood in September 2010 when over 200 mm of rain was recorded in 3 days in the nearby meteorological station at Škocjan with maximal precipitation rates exceeding 50 mm/h. Flood deposits show that the complete entrance shaft series was filled with water

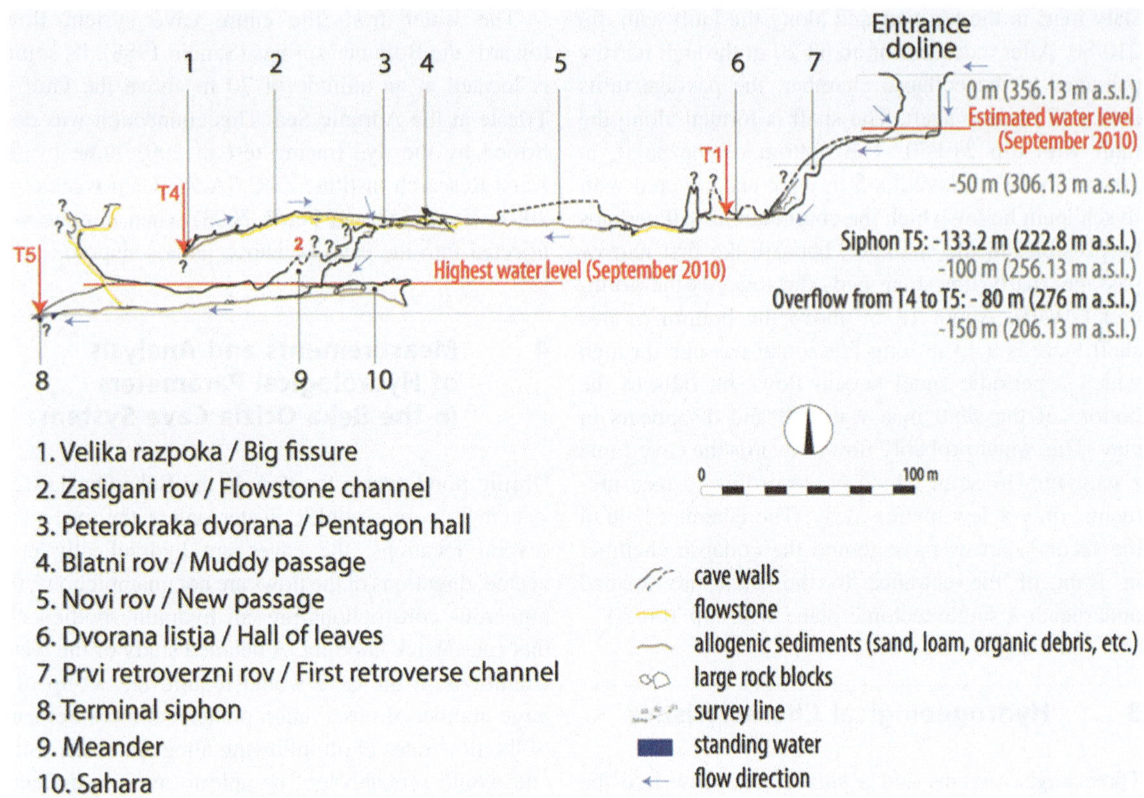

Fig. 13 Extended elevation of the cave Ocizeljska jama. The location of the observation points T 1, T 4 and T 5, and maximal water levels of the extreme flood event in September 2010 are marked in red

(Fig. 13) and the height of the water column at T 1 exceeded 30 m. After the station was set up in 2012, no similar event has been recorded since.

T 2 Maletova jama s slapom The cave is a sink where allogenic stream enters the cave through an 8 m high waterfall. The terminal siphon is filled with sediments, however, very close to a passage which stretches from the nearby Jama z naravnim mostom. The logger was fixed into an overflow pool about 10 m above the level of the sump. The known geometry of the overflow allows a crude estimation of flow rates (see Fig. 14).

T 3 Jama z naravnim mostom The cave is the main conduct of low and medium waters. The main channel is fully flooded at several locations during the flood events. The cave is connected to Jurjeva jama and Blažev spodmol and almost connected to Maletova jama s slapom. The point T 3 is also in an overflow pool with a 2 m high cascade above. The locations of the measurement points T 2 and T 3 enable rough estimation of flow rates during the flood events (see Fig. 14).

T 4 Zasigani rov (Flowstone Channel) in Ocizeljska jama The main junction in Ocizeljska jama is Peterokraka dvorana (Pentagon Hall). The water continues along the meander with a series of drops towards the terminal sump at the end of the passage Velika razpoka. Another channel, parallel to the passage Velika razpoka, initially ascends from the Peterokraka dvorana, then bends towards the southeast and descends down from the level 277–254 m where sediment deposits prevent its further exploration. The sediment (flysch sand with organic debris and pieces of garbage) clearly indicates the vicinity of the surface sinking stream. The loggers were fixed 2 m above the bottom to prevent them being buried by the sediments which are redistributed due to the upwelling water during the flood event. The positions of T 4 and T 5 points are marked in Fig. 13.

T 5 A sump lake at the end of the passage Rov velike razpoke The deepest point of the system is a lake at the end of the passage Rov velike razpoke in

Fig. 14 *Top left* Schematic presentation of the observation points T 2 and T 3. *Top right* The flowstone step below the overflow pool at T 3. *Bottom left* The temperature and pressure sensor with built-in datalogger ready to be attached to the stream bank. *Bottom right* A waterfall entering the cave Maletova jama s slapom

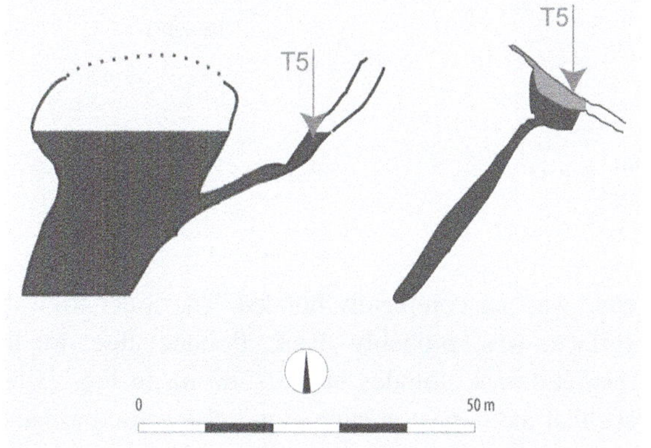

Fig. 15 The cross-section and ground plan of the terminal lake of the cave Ocizeljska jama with the observation point T 5

Ocizeljska jama. The lake is located at an altitude of 220 m. The divers explored it in 1990 and found a chamber with open water surface on the other side of the sump but no clear continuation. The lake is up to 24 m deep. Figure 15 shows the cross-section and ground plan of the lake.

5 Results

The extreme flood event of 2010 Between 17 and 19 September 2010, over 200 mm of rain fell in a series of intense events with rain intensity reaching over 50 mm/h. At that time T 5 was the only observation point in the cave system. Based on the occurrence of

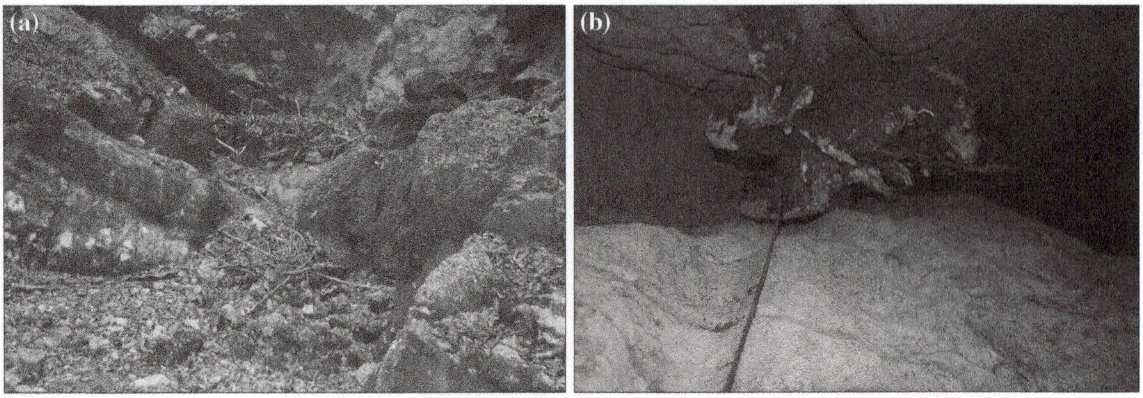

Fig. 16 Flood deposits in the entrance collapse doline of Ocizeljska jama (**a**); a small stump stuck in the passage of Blažev spodmol next to the exit to the collapse doline of Ocizeljska jama (**b**) (recorded on 6 November 2010)

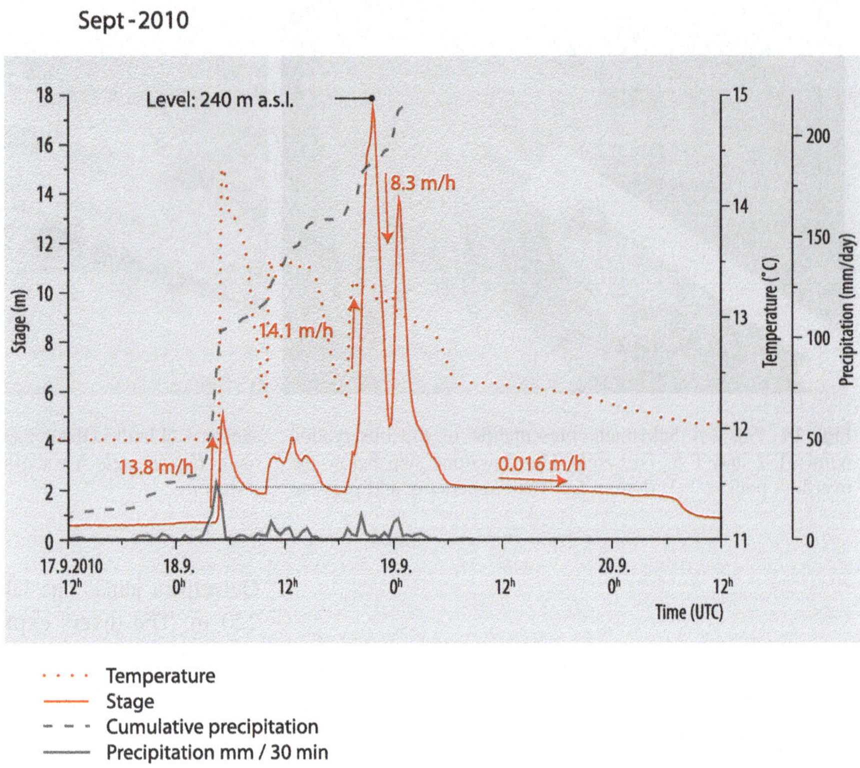

Fig. 17 The water level and temperature at T 5 and precipitation at the Škocjan meteorological station during the extreme flood event of September 2010

flood deposits before and after the event, it was evident that the entrance collapse doline of Ocizeljska jama was flooded over 5 m high (Figs. 16a, b and 17). The entrance series of shafts was surely completely flooded. At the bottom of the cave (T 5), however, the water level rose up by 17 m, to about 240 m above sea level, therefore, much of the deeper part of the cave was not completely flooded. The water from the surface was probably back flooded due to the restrictions at altitudes of 270–280 m. In Fig. 17 we see that the largest precipitation pulse caused a rise of about 4 m, however, later smaller events gave rise to subsequent peaks reaching up to 17 m above the initial level.

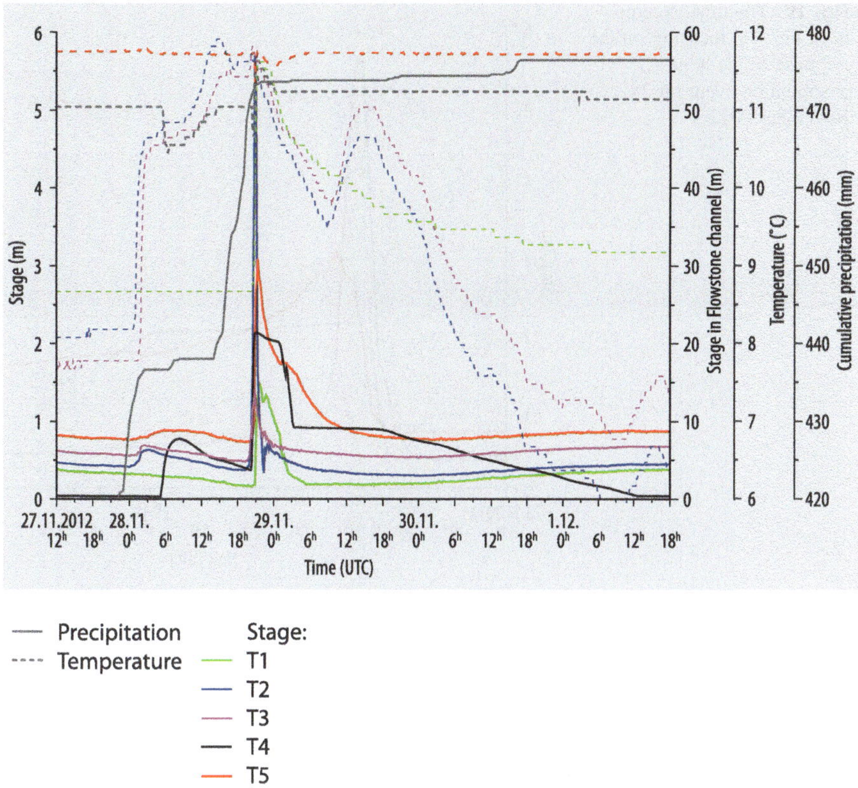

Fig. 18 The flood event at the end of November 2012 as recorded at all stations. The level at T 3 is shown in different scale on the right axis

Flood events between October 2012 and March 2013 The reaction of the system to a small precipitation event is practically unimportant. The flow along open surface streams increases, some intermittent sumps appear, but no substantial back flooding is observed. Figure 18 shows a larger flood event between 27 November and 1 December 2012. About 50 mm of rain was recorded in the nearby meteorological station at Škocjan, with maximal rain intensities reaching 10 mm/h–20 mm/h. The first rain pulse is barely recorded: we see a similar temperature dynamics on the points T 2 and T 3. What is interesting is a delayed stage response at T 4 where water rises for 7 m with a 6 h delay and then slowly recedes.

A vigorous response of stage at all stations follows the second rain pulse, when 30 mm of precipitation fell on 28 November. Details are shown in Fig. 19. At the point T 4, the water rises till 21 m, with a rate of up to 32 m/h. There, the maximal rise is limited by the position of overflow into Peterokraka dvorana. Deflections in the temperature signal at T 4 and T 5 point that a travel time between two points is in the range of 10–20 min.

Once the overflow is reached, the stage at T 4 remains relatively constant. As the recharge is reduced, it drops down to 9 m where it remains stagnant for another 12 h. The sump empties with slow recession of about 0.2 m/h.

When the flood water from T 4 reaches T 5, water rises to about 2 m and later drops along with decrease in the recharge.

The observation points T 2, T 3 and T 4 are all close to the sinking points, therefore they show similar temperature dynamics. The level at T 2 rises for 1.5 m. This is the only event where T 2 was flooded. At T 3 the level rises to 5 m, pointing to back flooding, due to the limited capacity of the outflow sump. The level at T 4 rises to 1 m. The maximal flow rate of 2 m^3/s at T 3 was estimated based on this level and the rough geometry of the overflow.

Figure 20 shows an interpretation of the flood event between the points T 4 and T 5. The recharge to T 4 is yet unknown, most probably the water comes from some of

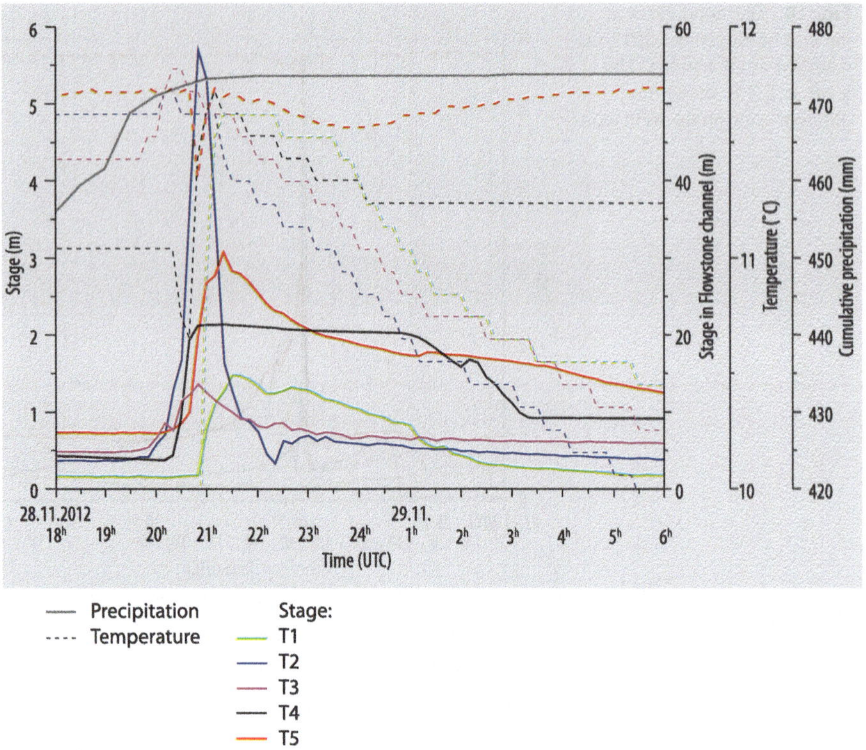

Fig. 19 The time excerpt from Fig. 18, focusing on the response to the largest precipitation event on 28 November 2012

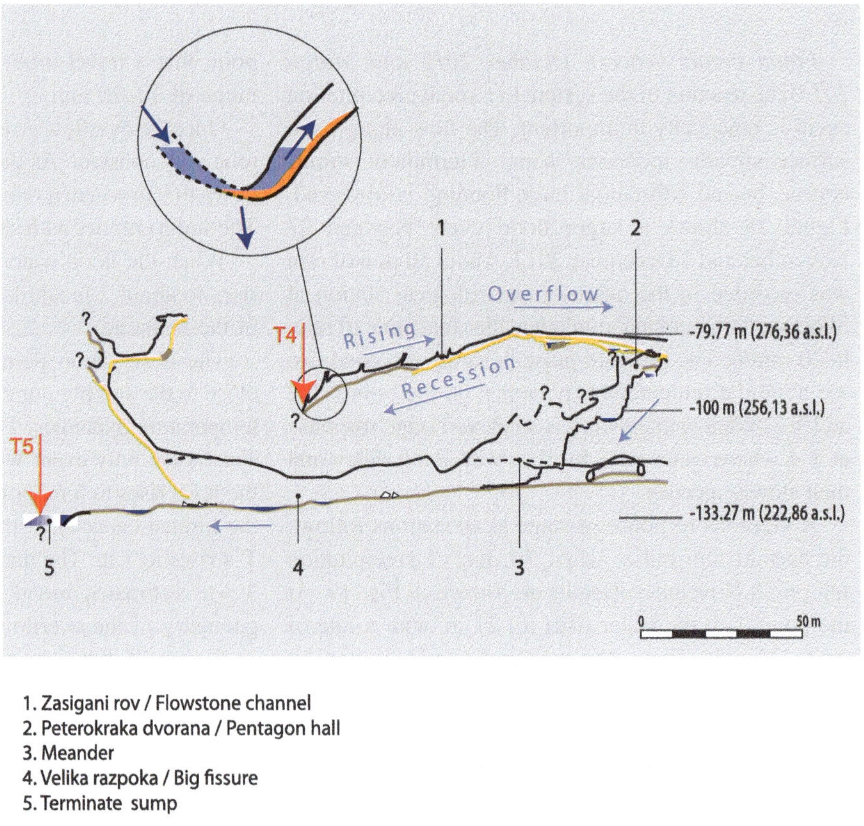

Fig. 20 Excerpt from the cross-section of the cave Ocizeljska jama, pointing to the overflow from the passage Zasigani rov (T 4) to the passage Rov velike razpoke and the terminal sump at T 5. The lower sketch shows the intermittent sump at T 4, as predicted from the measurements

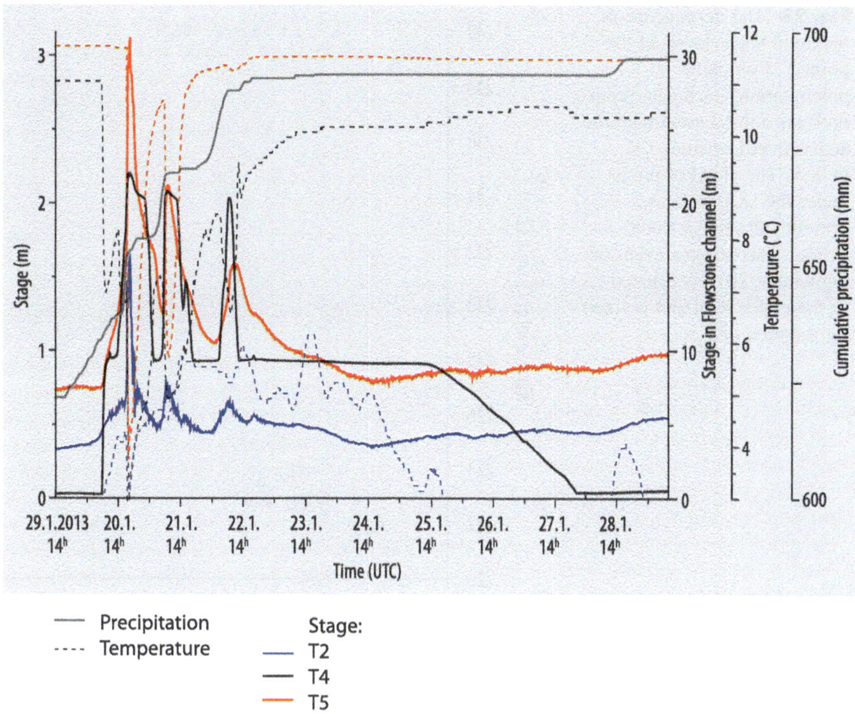

Fig. 21 Series of flood events in January 2013 recorded at the points T 2, T 4 and T 5

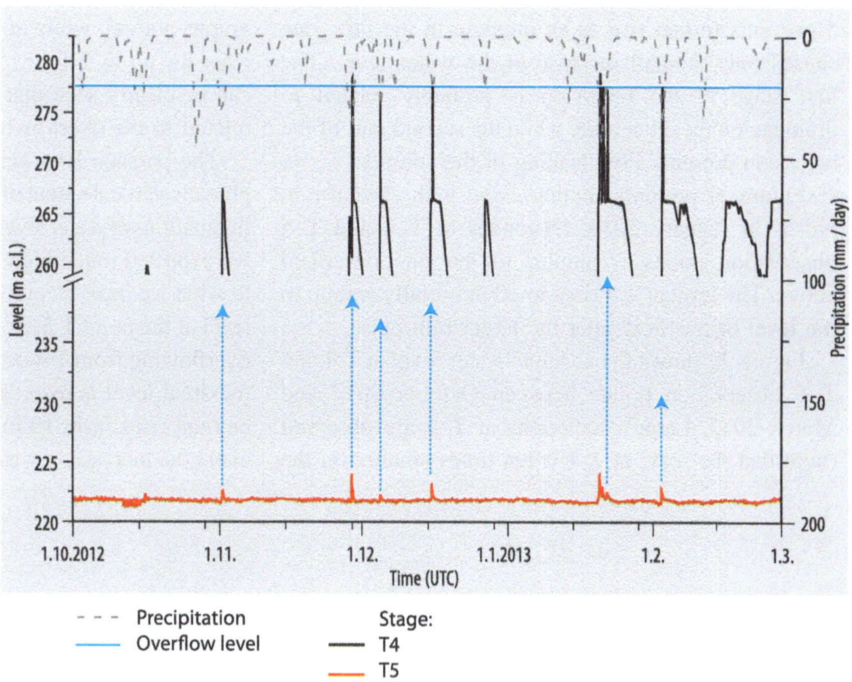

Fig. 22 The level of water surface (metres above sea level) at T 4 and T 5 observation points between October 2012 and March 2013. The dashed line shows daily amount of precipitation. The hydrogram at T 4 shows only values when the logger is submerged (i.e. 2 m above the bottom of the channel). Blue arrows indicate events when the overflowing from T 4 to T 5 occured

the side passages of the cave Jama z naravnim mostom. A candidate could be the shaft Vodnjak, which is poorly explored. According to flood deposits, the water comes relatively directly from the surface. Furthermore, the temperature at T 4 is similar to the temperatures at the points that are known to be close to the surface (T 1–T 3). Although the passage to the other side is blocked, we can draw some conclusions on it. There the water obviously rises at the same rate as in the Zasigani rov passage. The overflow towards the observation point T

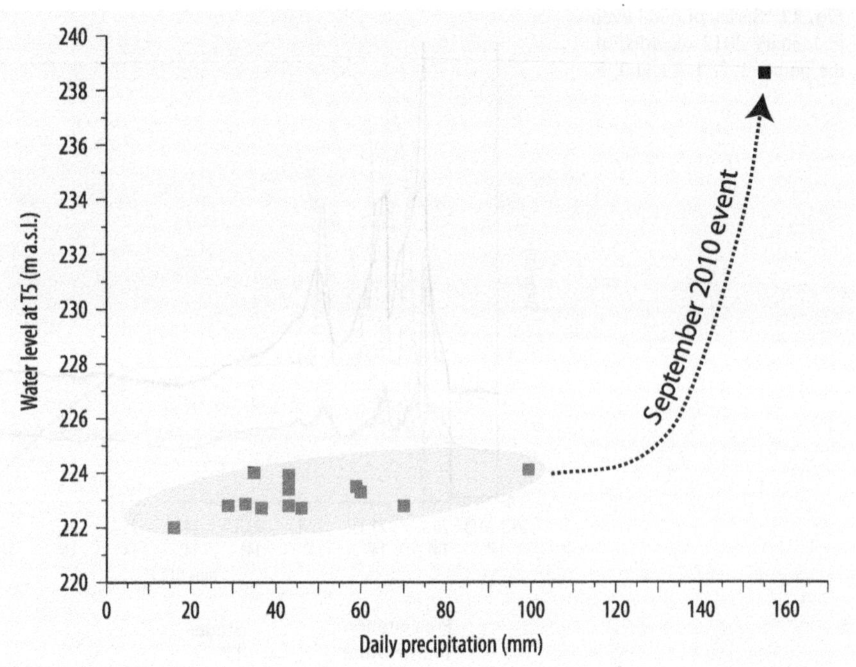

Fig. 23 The dependence of maximal water levels at the point T 5 on daily precipitation. Each data point presents a flood event with the active overflow from T 4 to T 5. The cloud of points under the shaded areas presents all regular flood events. The extreme event of September 2010 is isolated in the upper right and pointed by a dotted line

5 prevents further rise as an increase in the flow rate causes only a small increase in the water level. The first stage of the recession is probably related to drainage on the other side, while the second part of the recession presents slow leaking of the sump.

Figure 21 presents another event with several rain pulses in January 2013. Dynamics at T 4 and T 5 observation points is similar to the one described above. The level at T 4 rises to 10 m initially and up to the level of overflow after the larger rain pulse.

Figure 22 shows the absolute water level at T 4 and T 5 observation points between October 2012 and March 2013. Large oscillations at T 4 are observed (note that the scale at T 4 is ten times smaller on the graphs above), while at the outflow sump at T 5 water rises for up to 2 m, except for the extreme events. It can be clearly seen that all substantial rises at T 5 are related to the overflowing from T 4.

The passage Rov velike razpoke has a clearly epiphreatic vertical extent of up to 40 m, which indicate high fluctuations of water level during its formation. Therefore we expected much higher recent fluctuations compared to what we have recorded. Figure 23 shows the highest level at the point T 5, as recorded at all events when the overflowing from T 4 occurred. Only a weak increase in maximal level is recorded when the daily precipitation amount rises from 10 to 100 mm. Only at the extreme event the increase in water level is substantial.

Conclusion

Karst Research Institute ZRC SAZU is involved in individual projects related to the development and protection of the natural and cultural heritage of karst areas, regional planning, water supply systems, the construction of transportation infrastructure, etc. After two already published books (Karstology and Development Challenges on Karst I and II, ZRC Publishing) related to karst water, construction, tourism, ecology and protection, this book treats karst research as preceding studies of the railway construction on Slovenian karst. 43 % of Slovenian territory represents karst areas and more than 50 % of water supply comes from the karst. Slovenia is a country of the Classical Karst, from which the first international description of karst phenomena has been derived.

We are aware of the need for the continuous and effective communication of karstological knowledge to the wider social community, including through our participation in important and directly useful projects. By applying high-quality karstological research results from individual spheres of karstology and interdisciplinary studies, a foundation for sustainable planning of life in karst regions that will consider the natural and cultural characteristics and the vulnerability of the karst landscapes has been established. The proposed railway between the Northern Adriatic ports of Koper (Slovenia) and Trieste (Italy) and the interior of Slovenia on the 5th European Railway Corridor (Venice–Kiev) required extensive karstological planning of the route.

Karst is a result of the long-lasting evolution of the entire area, during which the surface and underground drainage paths changed, but relict caves were preserved in the karst. We focused our attention primarily on the location of larger segments of the surface, on the size and distribution of collapse dolines and unroofed caves which karst denudation had opened up in the surface.

With a geomorphological analysis of the karst and non-karst surface above the planned route we cannot precisely determine where and what kind of cavities exist in the underground, but it does help us to determine the zones and depths at which one can expect greater porosity or the location of important cavities.

The morphology of the karst indicates that today all the precipitation water from the area of the tunnels flows through the karst and does not create large cavities. Creviced corrosion shafts are possible anywhere on the tunnel route.

Combining the results of various research studies leads to the conclusion that the entire area is highly hollowed. It is believed that from 5 to 10 caves are likely to open up across 1 km of the route, and 15 caves in the separately described areas.

Therefore, the probability of the tunnels intersecting karst cavities in the above-mentioned areas is high. Cave passages, segments of relict cave networks, can be expected along the entire route at various levels. These passages can measure over 10 m in diameter. Coming across such passages is especially likely between the villages of Divača and Lokev where cave passages and networks, products of older speleogenetic stages, can be expected between the surface and

M. Knez et al. (eds.), *The Beka-Ocizla Cave System*, Cave and Karst Systems of the World,
DOI: 10.1007/978-3-319-04456-9_6, © Springer International Publishing Switzerland 2015

phreatic zone. The tunnels run across several contacts of limestone and flysch; there the probability of encountering cavities and oscillating groundwater is very high. The highly perforated zone behind the contact of flysch and limestone beneath the village of Beka should be pointed out. It can be expected with near certainty that karst cavities will be encountered there during the construction of the tunnel T 2. A similar level of cavernosity is expected beneath the Kastelec road-tunnel in the vicinity of the Brezno na Škrklovici shaft. In light of the research conducted in boreholes and during the construction of motorways, it has been deduced that the entire karst massif is highly karstified. There is a high probability of encountering shafts along the entire route which drain water to the level of the karst groundwater.

The **Beka-Ocizla cave system**, which is composed of six caves is formed on the contact karst between limestone and flysch in a contact depression. The tunnel T 2 will run in between the known passages of the cave network as these were deliberately avoided during the planning stage. However, it is highly probable that unknown segments of the cave will be encountered. Hence, the cave was once more thoroughly explored.

The entrance parts of the caves are developed in marly limestone but the main parts of the passages are in the Alveolina-Nummulites limestone with the 70/20 dip of the beds. The phreatic forms of certain passages of the Ocizeljska jama cave show that these parts of the cave had already developed under phreatic conditions and were later followed and transformed by sinking streams. The latter also formed new passages.

The most important geological structural elements which influenced the formation and development of the cave system and the orientations of its passages are: the contact between limestone and flysch in the NW–SE direction; the system of parallel fault planes with the dip of 220/50(80); these fault planes are the most important permeable structures and give the general orientation of the caves; the fissures with the dip of 310/80(90) are important for the orientation of shorter sections of the passages.

The majority of water that is flowing through the cave system is of allogenic origin. Therefore, the oscillation of discharge is high and ranges from less than one l/s to several m^3/s. The water flow in the system shows complex behaviour and has not yet been completely resolved. From the hydrological point of view, the most important cave in the Beka-Ocizla cave system is Ocizeljska jama, which was explored most thoroughly. Hydrological observation of water levels using dataloggers points out that during extreme events the upperparts of Ocizeljska jama are completely flooded, due to the back flooding caused by local restrictions, while the lower parts remain only partially flooded. The observations indicate that, although the passage density is high, there are still unknown channels with important amounts of sinking water; such is the recharge channel on the other side of the Zasigani rov passage. The measurements clearly show that the flood water which raises the water level at the terminate sump arrives from Zasigani rov, which enables overflow.

Caves and other **subterranean** habitats are an important part of both the terrestrial and aquatic ecosystems in karst terrains. The Dinaric karst, and especially its Slovenian part, is a global hotspot for subterranean biodiversity. The subterranean fauna of Slovenia is among the richest in the world. In the study area, the fauna from the infiltration zone was sampled. Boreholes in the karst underground offer a unique opportunity of direct access to the habitat developed in the unsaturated zone. In the boreholes aquatic or terrestrial fauna that lives in the system of cracks and fissures below the surface can be sampled directly. The karst underground domain, which at first glance seems to be averse to the existence and development of life, is inhabited by a large number of different organisms. The sampling of selected subterranean habitats provides insight into the rich subterranean biodiversity. Subterranean species can be used as bioindicators to help assess the overall ecological status of subterranean habitats.

By understanding the basic hydrogeological conditions and, above all, on the basis of the three tracer experiments conducted, we have been able to infer directions and characteristics of the groundwater flow from the planned railway route and endangerment of karst water sources within its impact area. The Rižana spring must be especially pointed out as a regionally important water source which has been captured for the water supply of towns on the Slovenian coast. Potential contamination from the southern part of the route (T 2 tunnel), which is protected with the Decree on determining the drinking water protection area for the aquifers of Rižana as a part of a broader water protection area, would appear in the Rižana spring during a high water level after only 4 days. It could be detected about a day sooner in the Osapska reka spring and, if a suitable observation network were organized, the danger of the Rižana spring being contaminated could be predicted as well. As the water level drops, the transfer of substances slows down somewhat, whereas the portion of the outflow towards the Rižana spring increases. From the remaining part of the planned T 2 tunnel waters flow towards the Rižana and Osapska reka springs more slowly and to a smaller degree.

The waters from the northern part of the T 2 tunnel mostly flow to the permanent spring Boljunec/Bagnoli della Rosandra in Italy. Potential contamination, coming from the location, would appear in the spring in high concentrations after approximately 4 days during a low water level, and somewhat sooner during a high water level. A good connection between this spring and the southern part of the T 1 tunnel has also been proved. As regards the flysch area, where the route runs across the surface, it has been predicted that water will flow into the surface flow of the Glinščica River. From the area of the T 1 tunnel the groundwater flows primarily towards the Timava spring and to a very small degree also towards other, smaller springs in the Gulf of Trieste in Italy. The velocities of the underground flow discovered by the tracer experiments show that during a low water level contamination from the area of the route would appear in the Timava springs after about 1 month, and within 1 week during a high water level. Since the Timava spring has a very large catchment area and the portion of water contributed from the area of the route is relatively small, any exceeding of the allowed concentrations could be detected in the Timava River only after large enough quantities of harmful substances were spilt, due to the high level of dilution.

The endangerment of karst water sources is therefore great, since even small quantities of harmful substances that would appear in the springs in very low concentrations can have harmful long-term effects. Significant quantities of these substances could accumulate when flowing through the vadose zone and would afterwards be gradually washed out with precipitation. In the event of several consecutive contamination events the quantity of accumulated substances would increase and the washing out period would last significantly longer. In addition to the quantity of contamination, the type of contamination is also very important. Experience thus far shows that washing out of the substances that are not soluble in water (e.g. oil derivates) is significantly different and above all lengthier than in the case of soluble types of contamination. Particularly harmful are substances that are slowly degradable and whose degradation products are hazardous to humans. When it comes to contamination on the karst surface, we must be aware of the fact that once the contamination passes through the layer of soil and enters the fissured karts rock, rehabilitation is no longer possible. As for the tunnels that are already located within karst aquifers, a negative impact can be prevented only by building watertight structures.

Counts of naturally present cultivable bacterial indicator groups in the Boljunec, Rižana and Osapska reka springs were used to assess the quality of the springs, to use **microbiota** as an "indigenous tracer", and to correlate microbial counts with the environmental variables. In the period of the winter of 2010/2011, high concentrations of total cultivable bacteria were occasionally detected, and in addition, in all three springs. Discharge was positively correlated with the concentration of bacteria in the Boljunec and Rižana springs, but not in the Osapska reka spring, which, in contrast to the other two, is not active all year round. Correlations between pH, SEC and temperature, and bacterial indicators can be attributed to the specific local conditions at each spring. The Kolmogorov-Smirnov test showed a similar distribution of bacterial indicators in all karst springs which may indicate a comparable situation in the recharge area for all three springs with respect to the biological load.

Once again, the effort to **develop integral karstology** that uses select approaches for getting to know and understand the uniform, three-dimensional karst landscape has proved to be the best choice. Under specific research conditions we have been able to get a good image of the karst across which the railway will run. It is true that we could not precisely determine the locations of caves that will be opened during the construction and all the paths of waters that shape and connect the karst, but we could clearly predict what is to be expected during the construction and what this construction should be like.

This will make the construction more successful; it will be possible to protect more of the newly discovered karst phenomena, which are an important part of our natural heritage; moreover, the waters that are also important for supply will be protected more efficiently.

The monitoring of any potential negative impacts of the construction and operation of the railway line will be possible only if a comprehensive analysis of the current status of the quantities and qualities of the above-mentioned springs during various hydrological conditions is performed prior to construction. During construction and operation we will have to carry out karstological monitoring, which will aid us in quickly overcoming obstacles during construction and identifying natural heritage, and a more thorough monitoring of the above-mentioned springs, taking into account the special characteristics of karst aquifers and the sampling dynamics, which will be suitably adapted to the precipitation and hydrological conditions.

Supplements

Supplement 1 Directions of underground water flow: for individual sections of the planned railway route (marked with broken lines in different colours) the most probable directions of the underground water flow are marked with arrows in different colours (dark blue for the Timava spring, light blue for the Boljunec spring, yellow for the Glinščica surface stream, green for the Rižana spring and orange for the Osapska reka spring); a solid line represents the main direction (a thicker line means larger share of flow in this direction) and a dotted line the secondary direction proved by the tracer tests; a dashed line represents possible connection which was not proved by the tracer test.

Supplement 2 Along the route of the railway from Črni Kal towards north to Divača, .avi file. Black: railway, red: depth of different types of caves, blue: boreholes (technical design S. Glažar).

Supplement 3 Along the route of the railway from Divača towards south to Črni Kal, .avi file. Black: railway, red: depth of different types of caves, blue: boreholes (technical design S. Glažar).

Supplement 4 Around the Beka-Ocizla Cave System, .avi file. Black: railway, red: depth of different types of caves, blue: boreholes (technical design S. Glažar).

M. Knez et al. (eds.), *The Beka-Ocizla Cave System*, Cave and Karst Systems of the World,
DOI: 10.1007/978-3-319-04456-9, © Springer International Publishing Switzerland 2015

References

Bosák P, Mihevc A, Pruner P, Melka K, Venhodová D, Langrová A (1999) Cave fill in the Černotiče Quarry, SW Slovenia: paleomagnetic, mineralogical and geochemical study. Acta Carsologica 28(2):15–39

Bosák P, Knez M, Otrubová D, Pruner P, Slabe T, Venhodová D (2000) Paleomagnetic research of fossil cave in the highway construction at Kozina (Slovenia). Acta Carsologica 29(2):15–33

Bosák P, Mihevc A, Pruner P (2004) Geomorphological evolution of the Podgorski Karst, SW Slovenia: contribution of magnetostratigraphic research of the Črnotiče II site with *Marifugia* sp. Acta Carsologica 33(1):175–204

Bratoš K, Sancin S (1984) Ocizelska pečina. Naše jame 26:89–93

Bressan G, Snidarcig A, Venturini C (1998) Present state of tectonic stress of the Friuli area (eastern Southern Alps). Tectonophysics 292(3/4):211–227

Caves Registry. Karst Research Institute ZRC SAZU, Postojna, Speleological Association of Slovenia, Ljubljana

Cività M, Cucchi F, Eusebio A, Garavoglia S, Maranzana F, Vigna B (1995) The Timavo hydrogeologic system: an important reservoir of supplementary water resources to be reclaimed and protected. Acta Carsologica 24:169–186

Culver DC, Pipan T (2013) Subterranean ecosystems. In: Levin SA (ed) Encyclopedia of biodiversity, 2nd edn. Academic Press, Waltham, pp 49–62

Culver DC, Pipan T, Gottstein S (2006) Hypotelminorheic—a unique freshwater habitat. Subterr Biol 4:1–8

Decree on determining the drinking water protection area for the aquifers of Rižana (2008). Official Gazette of the Republic of Slovenia 49/2008, Ljubljana

Gabrovšek F, Peric B (2006) Monitoring the flood pulses in the epiphreatic zone of karst aquifers: the case of Reka river system, karst plateau, SW Slovenia. Acta Carsologica 35 (1):35–45

Gabrovšek F, Knez M, Kogovšek J, Mihevc A, Mulec J, Petrič M, Slabe T, Šebela S, Zupan Hajna N (2001) Krasoslovna študija področja, na katerem se načrtuje gradnja drugega tira železniške proge Divača–Koper (varianta I/3). Elaborate of the Karst Research Institute ZRC SAZU, Postojna, p 81

Gams I (2004) Kras v Sloveniji v prostoru in času (Summary: Karst in Slovenia in space and time), 2nd edn. Založba ZRC, Ljubljana

Geldreich EE (1970) Applying bacteriological parameters to recreational water quality. J Am Water Works Assoc 62 (2):113–120

Habič P (1985) Vodna gladina v Notranjskem in Primorskem krasu Slovenije. Acta Carsologica 13:37–77

Hammer O, Harper DAT, Ryan PD (2001) PAST: paleontological statistics software package for education and data analysis. Palaeontol Electron 4:1–9

Jurkovšek B, Toman M, Ogorelec B, Šribar L, Drobne K, Poljak M, Šribar Lj (1996) Formacijska geološka karta južnega dela Tržaško-Komenske planote 1:50.000: kredne in paleogenske karbonatne kamnine. (Geological map of the southern part of the Trieste-Komen plateau 1:50,000: cretaceous and paleogene carbonate rocks. Inštitut za geologijo, geotehniko in geofiziko, Ljubljana, p 143

Knez M (1996) Vpliv lezik na razvoj kraških jam. Primer Velike doline, Škocjanske jame. (Summary: the bedding-plane impact on development of karst caves. An example of Velika Dolina, Škocjanske Jame Caves. Zbirka ZRC 14, ZRC SAZU, Ljubljana, p 186

Knez M, Slabe T (1999) Unroofed caves and recognizing them in karst relief (discovered during motorway construction at Kozina, South Slovenia). Acta Carsologica 28(2):103–112

Knez M, Slabe T (2000) Jame brez stropa so pomembna oblika na kraškem površju: s krasoslovnega nadzora gradnje avtocest na krasu. (Denuded caves as important feature on karst surface; recognition from supervision of motorway construction on karst). In: Gostinčar A (ed) Zbornik povzetkov referatov s 5. slovenskega kongresa o cestah in prometu. Družba za raziskave v cestni in prometni stroki Slovenije, Ljubljana, p 29

Knez M, Slabe T (2001) Karstology and expressway construction. In: Proceedings of the 14th IRF Road World Congress, Paris

Knez M, Slabe T (2002) Unroofed caves are an important feature of karst surfaces: examples from the classical karst. Z Geomorphol 46(2):181–191

Knez M, Slabe T (2004a) Karstology and the opening of caves during motorway construction in the karst region of Slovenia. Int J Spel 31(1/4):159–168

Knez M, Slabe T (2004b) Highways on karst. In: Gunn J (ed) Encyclopedia of caves and karst science. Fitzroy Dearborn, New York, pp 419–420

M. Knez et al. (eds.), *The Beka-Ocizla Cave System*, Cave and Karst Systems of the World,
DOI: 10.1007/978-3-319-04456-9, © Springer International Publishing Switzerland 2015

Knez M, Slabe T (2005) Caves and sinkholes in motorway construction, Slovenia. In: Waltham T, Bell F, Culshaw M (eds) Sinkholes and subsidence. Karst and cavernous rocks in engineering and construction. Springer, Praxis, pp 283–288

Knez M, Slabe T (2006a) Krasoslovne raziskave pri gradnji avtocest preko slovenskega krasa (Summary: karstological research during the construction of motorways crossing the Slovene Karst). Annales 16(2):259–266

Knez M, Slabe T (2006b) Dolenjska subsoil stone forests and other karst phenomena discovered during the construction of the Hrastje-Lešnica motorway section (Slovenia). Acta Carsologica 35(2):103–109

Knez M, Slabe T (eds) (2007) Kraški pojavi, razkriti med gradnjo slovenskih avtocest (with summary). Carsologica 7, Založba ZRC, Ljubljana, p 250

Knez M, Slabe T (2009) Caves in breccia and flysch below Mount Nanos in the Vipava Valley (Slovenia). Annales 19 (1):85–94

Knez M, Slabe T (2010) Planning and constructing motorways and railroads crossing Classical Karst in Slovenia. In: Proceedings of the first international conference on road and rail infrastructure CETRA, Opatija, Croatia, May 2010. Department of transportation, Faculty of civil engineering, University of Zagreb, Zagreb, pp 837–843

Knez M, Slabe T (2011) Young karst processes in breccia and flysch (Mount Nanos, Slovenia). Presentaciones del II Congreso de América Central sobre Karst y Espeleología, I Congreso de Terrenos Kársticos de Guatemala (CD). Cobán, Guatemala

Knez M, Slabe T (2012a) Karstology in motorway construction on classical karst. In: Zorn M, Ciglič R, Perko D (eds) Geographical tidbits from Slovenia: special issue on the occasion of the 32nd International Geographical Congress in Cologne. Geografski vestnik 84(1):77–86

Knez M, Slabe T (2012b) Kraški pojavi, razkriti med gradnjo avtoceste na nizkem krasu Dolenjske. In: Paper presented at the 11th Slovenian road congress, Portorož, October 2012

Knez M, Slabe T (2012c) Expressway construction on young karst in breccia (Vipava Valley, Slovenia). In: Lakušić S (ed) Road and rail infrastructure II. Proceedings of 2nd international conference on road and rail infrastructure CETRA, Dubrovnik, Croatia, May 2012. Department of transportation, Faculty of civil engineering, University of Zagreb, Zagreb, pp 773–779

Knez M, Slabe T (2012d) Planning, research and karstological monitoring of expressways crossing classical Karst (Slovenia). Creative Educ 3(7B):39–42

Knez M, Šebela S (1994) Novo odkriti kraški pojavi na trasi avtomobilske ceste pri Divači. Naše jame 36:102

Knez M, Kranjc A, Otoničar B, Slabe T, Svetličič S (1994) Posledice izlitja nafte pri Kozini. Ujma 9:74–80

Knez M, Otoničar B, Slabe T (2003) Subcutaneous stone forest (Trebnje, Central Slovenia). Acta Carsologica 32(1):29–38

Knez M, Slabe T, Šebela S (2004a) Karstification of the aquifer discovered during the construction of the expressway between Klanec and Črni Kal, Classical Karst. Acta Carsologica 33(1):205–217

Knez M, Slabe T, Šebela S (2004b) Karst uncovered during Bič-Korenitka motorway construction (Dolenjska, Slovenija). Acta Carsologica 33(2):75–89

Knez M, Slabe T, Šebela S, Gabrovšek F (2008) The largest karst caves discovered in a tunnel during motorway construction in Slovenia's Classical Karst (Kras). Environ Geol 54(4):711–718

Kogovšek J (1993) Kakšna je sestava voda, ki odtekajo z naših cest (summary: water composition flowing off our roads). Ujma 7:67–69

Kogovšek J (1995a) Izlitja nevarnih snovi ogrožajo kraško vodo. Onesnaženje Rižane oktobra 1994 zaradi izlitja plinskega olja ob prometni nesreči v Obrovu. Annales 5(7):141–148

Kogovšek J (1995b) Podrobno spremljanje kvalitete vode, odtekajoče z avtoceste in njen vpliv na kraško vodo. (Detailed monitoring of the quality of the water that runs off the motorway and its impact on karst water). Annales 5/7:149–154

Kogovšek J (1997) Pollution transport in the vadose zone. Karst waters and environmental impacts. In: Gunay G and Johnson AI (eds) Proceedings of the 5th international symposium and field seminar on waters and environmental impacts, Antalya, Turkey, September 1995. Balkema, Rotterdam, pp 161–165

Kogovšek J (2000) Ugotavljanje načina pretakanja in prenosa snovi s sledilnim poskusom v naravnih razmerah. Annales 19:113–142

Kogovšek J (2010) Characteristics of percolation through the karst vadose zone. Carsologica 10, Založba ZRC/ZRC Publishing, Ljubljana, p 168

Kogovšek J, Petrič M (2003) Tracing tests as a tool for the estimation of the possible impacts of human activities on karst waters—examples from Slovenia. RMZ-Mater Geoenviron 50(1):161–164

Kogovšek J, Petrič M (2004) Advantages of longer-term tracing—three case studies from Slovenia. Environ Geol 47:76–83

Kogovšek J, Petrič M (2007) Directions and dynamics of flow and transport of contaminants from the landfill near Sežana. Acta Carsologica 36(3):413–424

Kogovšek J, Slabe T, Šebela S (1997) Motorways in Karst (Slovenia). In: Byerly DW (ed) Proceedings and a fieldtrip excursion guide of the 48th highway geology symposium Knoxvill, Tennessee, May 1997. University of Tennessee, Department of Geological Sciences, pp 49–55

Kolbezen M, Pristov J (1998) Površinski vodotoki in vodna bilanca Slovenije. (Surface streams and water balance of Slovenia). Ministrstvo za okolje in prostor, Hidrometeorološki zavod Republike Slovenije, Ljubljana, p 29

Košir A (2003) Litostratigrafska revizija zgornje krede in paleogena v jugozahodni Sloveniji. Geološki zbornik 17:92–98

Košir A, Otoničar B (2001) The evolution of upper cretaceous and paleogene synorogenic carbonate platforms in NW Dinaric foreland basin. In: Dragičević I, Velić I (eds) Abstracts of the 1st scientific meeting on carbonate platform or carbonate platforms of Dinarides, Zagreb, Croatia, October 2001. Rudarsko-geološko-naftni fakultet, Prirodoslovno-matemetički fakultet, Institut za geološka istraživanja i Hrvatsko geološko društvo, Zagreb, pp 62–63

Kranjc A (ed) (1997) Kras: Slovene Classical Karst. Založba ZRC, Ljubljana, p 254

Krivic P, Bricelj M, Trišič N, Zupan M (1987) Sledenje podzemnih vod v zaledju izvira Rižane (summary: water tracing in the Rižana spring ground water basin). Acta Carsologica 16:83–104

Krivic P, Bricelj M, Zupan M (1989) Podzemne vodne zveze na področju Čičarije in osrednjega dela Istre (summary: underground water connections in Čičarija region and in Middle Istria). Acta Carsologica 18:265–295
Latella L, Sbordoni V (2002) Fauna delle Grotte. In: Minelli A, Chemini C, Argano R, Ruffo S (eds) La fauna in Italia. Touring Club Italiano and Ministero dell'ambiente e della tutela del territorio e del mare, Roma, pp 339–358
Marton E, Drobne K, Cimerman F, Ćosović V, Košir A (1995) Paleomagnetism of latest Maastrichtian through Oligocene rocks in Istria (Croatia), the Karst region, and S of the Sava fault (Slovenia). In: Vlahović I, Velić I, Šparica M (eds) Proceedings of the 1st Croatian geological congress, Opatija, October 1995. Institut za geološka istraživanja and Hrvatsko geološko društvo, pp 355–360
Mihevc A (1991) Morfološke značilnosti ponornega kontaktnega krasa; izbrani primeri s slovenskega krasa. Magistrska naloga, Univerza v Ljubljani
Mihevc A (1996) Brezstropa jama pri Povirju. Naše jame 38:65–75
Mihevc A (1999) The caves and the karst surface—case study from Kras, Slovenia. In: Karst 99: colloque européen: des paysages du karst au géosystème karstique: dinamiques, structures, et enregistrement karstiques. Études de géographie physique, suppl 28, Aix-en-Provence, pp 141–144
Mihevc A (2000) The fossilized tubes from the roofless cave—probably the oldest known remains of the cave worm Marifugia (Annelida: Polychaeta). Acta Carsologica 29 (2):261–270
Mihevc A (2001) Geomorfološko kartiranje na delu HC Razdrto-Vipava (Rebernice), ki poteka v območju krajinskega parka. Založba ZRC, Ljubljana, p 49
Mihevc A, Zupan Hajna N (1996) Clastic sediments from dolines and caves found during the construction of the motorway near Divača, on the Classical Karst. Acta Carsologica 25:169–191
Mihevc A, Slabe T, Šebela S (1998) Denuded caves—an inherited element in the karst morphology: the case from Kras. Acta Carsologica 27(1):165–174
Moe CL (1997) Waterborne transmission of infectous agents. In: Hurst CJ, Knudsen GR, McInerney MJ, Stetzenbach LD, Walter MV (eds) Manual of environmental microbiology. American Society for Microbiology, Washington, pp 136–152
Mosetti F (1989) Problemi di marcatura delle acque. Carsismo e idrologia carsica nel Friuli-Venezia Giulia. Quaderni ETP: rivista di limnologia 17:125–152
Mulec J, Krištůfek V, Chroňáková A (2012a) Comparative microbial sampling from eutrophic caves in Slovenia and Slovakia using RIDA®COUNT test kits. Int J Spel 41(1):1–8
Mulec J, Krištůfek V, Chroňáková A (2012b) Monitoring of microbial indicator groups in caves through the use of RIDA®COUNT kits. Acta Carsologica 41(2/3):287–296
Novak D (1964/1965) Hidrogeologija območja Osapske reke. Vesnik 4–5/B:81–91 (Beograd)
Oarga A, Griessler Bulc T, Jenssen PD, Mulec J (2012) Monitoring of microbial indicator groups in organically heavily loaded wastewater treatment systems by using RIDA®COUNT kits. Fresenius Environ Bull 21 (12a):3886–3893
Otoničar B (2007) Upper cretaceous to paleogene forbulge unconformity associated with foreland basin evolution (Kras, Matarsko Podolje and Istria; SW Slovenia and NW Croatia). Acta Carsologica 36(1):101–120
Petrič M, Kogovšek J (2011) Assessment of the possible impact of the construction of the Divača-Koper rail-way line on the quality of karst waters. In: Prelovšek M, Zupan Hajna N (eds) Pressures and protection of the underground karst: cases from Slovenia and Croatia. Karst Research Institute ZRC SAZU, Postojna, pp 138–146
Pipan T (2005) Epikarst–a promising habitat. Carsologica 5, ZRC Publishing, Ljubljana, p 101
Pipan T, Culver DC (2012) Shallow subterranean habitats. In: White WB, Culver DC (eds) Encyclopedia of caves, 2nd edn. Academic/Elsevier Press, Amsterdam, pp 683–690
Placer L (1981) Geološka zgradba jugozahodne Slovenije. Geologija 24(1):27–60
Placer L (1998) Contribution to the macrotectonic subdivision of the border region between Southern Alps and External Dinarides. Geologija 41:223–255
Placer L (2005) Strukturne posebnosti severne Istre (summary: structural curiosity of the northern Istria). Geologija 48 (2):245–251
Placer L, Vrabec M, Celarc B (2010) The basis for understanding of the NW Dinarides and Istria Peninsula tectonics. Geologija 53(1):55–86
Pleničar M, Polšak A, Šikić D (1969) Osnovna geološka karta SFRJ, list Trst, 1:100,000. Savezni geološki zavod, Beograd
Pleničar M, Polšak A, Šikić D (1973) Tolmač k Osnovni geološki karti SFRJ, list Trst, 1:100.000. Savezni geološki zavod, Beograd
Polák P, Roth T, Enčev J, Kocourek K (2012) Mezinárodni expedice Kačna jama Reka exploration 2011. Speleóforum 31:72–79
Rižnar I, Koler B, Bavec M (2007) Recent activity of the regional geologic structures in western Slovenia. Geologija 50(1):111–120
Sancin S (1984) Pod Socerbsko planoto. Pulje pri Domju, p 19
Sancin S (1988) Recenti indagini sulle sorgenti di Bagnoli. Rassegna Fed Spel Triestina 1:35–38
Singleton P, Sainsbury D (2000) Dictionary of microbiology and molecular biology, 2nd edn. Wiley, Chichester
Slabe T (1995) Cave rocky relief and its speleogenetical significance. Zbirka ZRC 10, Založba ZRC, Ljubljana, p 128
Slabe T (1996) Karst features in the motorway section between Čebulovica and Dane. Acta Carsologica 25:221–240
Slabe T (1997a) Karst features discovered during motorway construction in Slovenia. Environ Geol 32(3):186–190
Slabe T (1997b) The caves in the motorway Dane-Fernetiči. Acta Carsologica 26(2):361–372
Slabe T (1998) Karst features discovered during motorway construction between Divača and Kozina. Acta Carsologica 27(2):105–113
Stampfli GM, Mosar J (1999) The making and becoming of Apulia. In: Gosso G, Jadoul F, Sella M, Spalla MI (eds) Mem Sci Geol Special Volume 51/1 Third Workshop on Alpine Geological Studies, Biella-Oropa September–October 1997. Università di Padova, pp 141–154
Stampfli GM, Mosar J, Marquer D, Marchant R, Baudin T, Borel G (1998) Subduction and obduction processes in the Swiss Alps. Tectonophysics 296(1–2):159–204
Šebela S, Mihevc A (1995) The problems of construction on karst—the examples from Slovenia. In: Beck BF, Pearson

FM (eds) Karst geohazards, engineering and environmental problems in karst terrain. Proceedings of the 5th multidisciplinary conference on dolines and engineering and environmental impacts on karst. A.A. Balkema, Rotterdam, pp 475–479

Šebela S, Mihevc A, Slabe T (1999) The vulnerability map of karst along highways in Slovenia. In: Beck BF, Pettit AJ, Herring JG (eds) Hydrogeology and engineering geology of dolines and karst. Proceedings of the 7th multidisciplinary conference on dolines and the engineering and environmental impacts on karst. A.A. Balkema, Rotterdam, pp 419–422

Šikić D, Pleničar M, Šparica M (1972) Osnovna geološka karta SFRJ, list Ilirska Bistrica, 1:100 000. Savezni geološki zavod, Beograd

Timeus G (1928) Nei misteri del mondo sotterraneo. Alpi Giulie 29(1):1–40

Turk I, Bavdek A, Vidrih-Perko V, Culiberg M, Šercelj A, Dirjec J, Pavlin P (1992) Acijev spodmol pri Petrinjah, Slovenija: dr. Francu Lebnu-Aciju za petinšestdesetletnico. Poročilo o raziskovanju paleolita, neolita in eneolita v Sloveniji 20:27–48

WHO (1997) Guidelines for drinking-water quality. Surveillance and control of community supplies, 2nd edn, vol 3. World Health Organization, Geneva

Zupan Hajna N (2004) The caves of the contact karst of Beka and Ocizla, the SW Slovenia. Acta Carsologica 33 (2):91–105

Zupan Hajna N, Mihevc A, Pruner P, Bosák P (2008) Palaeomagnetism and magnetostratigraphy of Karst sediments in Slovenia. Carsologica 8, Založba ZRC, Ljubljana, p 266

Zupan Hajna N, Mihevc A, Pruner P, Bosák P (2010) Palaeomagnetic research on karst sediments in Slovenia. Int J Spel 39(2):47–60